I0053507

New Methods of Celestial Mechanics

Jan Vrbik

Department of Mathematics
Brock University
Canada

Contents

Foreword i

Preface iii

Chapter 1. Linearized Kepler problem 1

Chapter 2. Iterative solution of perturbed problem 18

Chapter 3. Perturbing forces 34

Chapter 4. Solar system 45

Chapter 5. Oblateness perturbations 57

Chapter 6. Lunar problem 67

Chapter 7. Resonances 79

Chapter 8. Other Perturbations 103

References 112

Index 115

Foreword

This book will be valuable to students and experts alike who are interested in elegant and efficient mathematical methods for achieving quantitative understanding of satellite orbit perturbations. Until the 20th century, celestial mechanics was the primary domain for development of analytic and computational techniques in mathematics, from treatment of complex singularities by Cauchy to recognition of chaos by Poincare. Classical mechanics was then pushed to the background by the rise of quantum mechanics until the beginning of the space age in the 1950s, when celestial mechanics was revived as a foundation for the new field of astromechanics to control space flight and artificial satellites. With its narrow mission-oriented focus, NASA incorporated 19th century mathematics into the space program without recognizing the importance of sponsoring further development of mathematical and computational techniques. Nevertheless, development continued to be driven by intellectual forces. This book is a good example of one outcome. Let me place it in a historical context.

The singularity of the gravitational potential presented a persistent problem in orbital mechanics until 1920, when Italian mathematician Tullio Levi-Civita discovered a change of variables that regularized the singularity and linearized the Kepler problem by transforming it to the form of a harmonic oscillator. However, his technique employed complex variables in way that limited the result to planar motion, and he was at a loss to generalize it to 3D. Thus, it remained as a mathematical curiosity until the 1964 Oberwolfach conference on Mathematical Methods in Celestial Mechanics when, in an opening address in the grand tradition of David Hilbert, Swiss mathematician Eduard Stiefel posed the generalization of Levi-Civita regularization to 3D as one of the great unsolved problems of mathematics. Then in dramatic fashion at the afternoon session, an obscure young Finnish astronomer Paul Kustaanheimo clamored for the podium to announce that he had solved the problem.

Kustaanheimo employed idiosyncratic spinor methods to generalize Levi-Civita's approach [23]. Unsatisfied with that, Stiefel transformed it to a more conventional matrix formulation now generally known as the Kustaanheimo-Stiefel (KS) transformation. Recognizing its advantages for perturbation theory in celestial mechanics, Stiefel promoted it in research and publicized it in a book [32].

After I heard this story about Kustaanheimo from a Finnish colleague, I reformulated the KS transformation in terms of geometric algebra, which I found clarifies its geometric structure and physical significance [20]. Pleased with the result, I called the method Spinor Perturbation Theory and incorporated it in my mechanics book [21]. Soon thereafter, a researcher in the space program informed me that he had applied it with considerable success in the design of software to control artificial satellites, but that was never published or followed up by others.

As abundantly demonstrated in this book, Vrbik has adopted the method and pushed it to a new level of perfection with detailed applications, including some of the trickier problems in planetary celestial mechanics. The reception to this book may well determine if Spinor Perturbation Theory is ultimately placed among the celebrated achievements of celestial mechanics.

Vrbik tells me he chose to formulate his theory in terms of quaternions rather than geometric algebra because the former has an established tradition in celestial mechanics. On the other hand, geometric algebra enhances the value of quaternions by clearly integrating them with vector algebra and spinors, thus resolving a history of confusion and controversy and consolidating their place in the broader context of mathematical physics [16]. It is an easy and instructive exercise to relate the quaternion formulation in this book to the spinor formulation in mine. The serious student will be in for some surprises.

Dr. David Hestenes
Professor Emeritus of Physics
University of Arizona, Tempe
2002 Oersted Medal recipient

Preface

This book has been inspired by the 'Celestial Mechanics' chapter of [21]. Its purpose is to demonstrate how the perturbed Kepler problem can be elegantly formulated and solved using the power of quaternion algebra (to be seen as a special case of a more general, and universally applicable, Geometric Algebra). The solution covers, in a natural and rather routine manner, even the most difficult cases of resonant perturbations.

There are two technical reasons which enabled me to construct such a universal solution. Namely, I had to (i) modify the usual definition of auxiliary time s (making the clock 'tick' faster, as the satellite approaches the primary, in a way which reflects the third Kepler law), and (ii) dispense with the traditional 'gauge' (this came at the expense of complicating the resulting equation — yet, without it, one is bound to fail, as the history of this problem clearly demonstrates). Only then can one proceed to build a unique solution in a rather algebraic manner of matching coefficients of two Fourier-series expansions (see Chapter 2). The rest of the book demonstrates how relatively easily can the resulting formulas be applied to practically all perturbations encountered within our Solar system.

The book is organized as follows: Chapter 1 reformulates the Kepler problem in terms of the quaternion algebra, and solves its unperturbed version. Chapter 2 constructs a detailed solution of the perturbed version; this is, in a sense, the key development of the book — however, it can be skipped by readers who want to learn only how to apply the resulting formulas. Chapter 3 derives and discusses most of the common perturbing forces, and explains their origin. Each of the remaining chapters concentrates on one such perturbing force, starting with perturbations of planetary motion caused by the gravitational pull of other planets (Chapter 4). For artificial satellites, the most important perturbing force is due to the Earth's slightly flattened shape (and similar deviations from a perfect sphere), the consequences of which are discussed in Chapter 5. For the Moon, the main perturbing force is the Sun; due to its relative strength, the Moon's motion becomes quite complicated and difficult to compute with high accuracy (Chapter 6). Chapter 7 deals with an interesting (and also rather difficult) phenomenon of a resonant perturbing force (meaning that is oscillates with a frequency which is either an exact multiple, or an exact fraction, of the orbital frequency) — this is important for proper understanding of asteroid motion, and it also explains the formation of Kirkwood gaps (narrow regions in the asteroid belt, which are almost void of asteroids). The final chapter then deals, rather more quickly, with the remaining common perturbations, such as the relativistic correction and radiation pressure.

Finally I would like to acknowledge that, as with any significant project, things are rarely done entirely by oneself. To this end I would like to thank: Dr. Hestenes without whom this book would not have been possible and my son Paul Vrbik for his help and technical expertise in preparing the LaTeX version of this book.

CHAPTER 1

Linearized Kepler problem

Abstract

Our first objective is to reformulate the age-old problem of a satellite orbiting a primary (of a substantially bigger mass) by means of quaternion algebra.[1] In subsequent chapters, this will facilitate a relatively simple construction of the problem's iterative solution.

Here we also review some details of constructing the *unperturbed* solution.

1.1 QUATERNION ALGEBRA

We first briefly review the algebra's basic rules and definitions [19], [16].

1.1.1 Basics

Quaternion algebra has three imaginary units usually called $\mathfrak{i}$, $\mathfrak{j}$ and $\mathfrak{k}$, each of them squaring to -1. Under the algebra's multiplication (denoted $\circ$), any two of these anticommute (i.e. $\mathfrak{i}\circ\mathfrak{j} = -\mathfrak{j}\circ\mathfrak{i}$), furthermore, $\mathfrak{i}\circ\mathfrak{j} = \mathfrak{k}$, $\mathfrak{j}\circ\mathfrak{k} = \mathfrak{i}$ and $\mathfrak{k}\circ\mathfrak{i} = \mathfrak{j}$. Together with the ordinary (real) unit 1 which commutes with each of them, they constitute the algebra's basis. An element of this algebra can thus be written as

$$A + a_z\mathfrak{i} + a_y\mathfrak{j} + a_x\mathfrak{k} \equiv A + \mathbf{a} \tag{1.1}$$

and interpreted geometrically as a VECTOR $\mathbf{a}$ (a_x, a_y and a_z representing its x, y and z coordinates, respectively), appended to a SCALAR A (note the deliberate reversal of the components of $\mathbf{a}$).

Quaternion MULTIPLICATION thus amounts to

$$(A + \mathbf{a}) \circ (B + \mathbf{b}) = AB - \mathbf{a}\cdot\mathbf{b} + A\mathbf{b} + B\mathbf{a} - \mathbf{a}\times\mathbf{b} \tag{1.2}$$

which can be easily verified by expanding $(A + a_z\mathfrak{i} + a_y\mathfrak{j} + a_x\mathfrak{k}) \circ (B + b_z\mathfrak{i} + b_y\mathfrak{j} + b_x\mathfrak{k})$. Note that this multiplication is noncommutative, but remains associative. The latter property is easily confirmed for any selection of the algebra's basis elements, e.g. $(\mathfrak{i}\circ\mathfrak{j})\circ\mathfrak{k} = \mathfrak{i}\circ(\mathfrak{j}\circ\mathfrak{k})$, $(\mathfrak{i}\circ\mathfrak{j})\circ\mathfrak{i} = \mathfrak{i}\circ(\mathfrak{j}\circ\mathfrak{i})$, etc., implying the general result.

From now on, all our VECTORS will be understood to be *quaternions* with zero real component, e.g. $\mathbf{f} \equiv f_z\mathfrak{i} + f_y\mathfrak{j} + f_x\mathfrak{k}$. A product of two vectors (such as, for example, $\mathbf{g}\circ\mathbf{f}$) will always imply their *quaternion* multiplication. Note that, in this notation, $\mathbf{f}^2 \equiv \mathbf{f}\circ\mathbf{f} = -f^2$, where $f = \sqrt{f_x^2 + f_y^2 + f_z^2}$ is the usual LENGTH of $\mathbf{f}$ (alternately denoted $|\mathbf{f}|$). Also note that, for the usual dot product, we get

$$\mathbf{g}\cdot\mathbf{f} = -\frac{\mathbf{g}\circ\mathbf{f} + \mathbf{f}\circ\mathbf{g}}{2} \tag{1.3}$$

and, correspondingly

$$\mathbf{g}\times\mathbf{f} = -\frac{\mathbf{g}\circ\mathbf{f} - \mathbf{f}\circ\mathbf{g}}{2} \tag{1.4}$$

[1]For more details see [20], [21], [23], [24], [32]–[39].

Jan Vrbik

for the cross product.

For quaternion quantities, we use the 'blackboard' type (e.g. $\mathbb{A} \equiv A + a_z\mathfrak{i} + a_y\mathfrak{j} + a_x\mathfrak{k}$); $\overline{\mathbb{A}}$ denotes the corresponding quaternion CONJUGATE (changing the sign of the $\mathfrak{i}$, $\mathfrak{j}$, and $\mathfrak{k}$-components, i.e. the vector part, of $\mathbb{A}$). One can easily verify that

$$\overline{\mathbb{A} \circ \mathbb{B}} = \overline{\mathbb{B}} \circ \overline{\mathbb{A}} \tag{1.5}$$

In general, the MAGNITUDE $|\mathbb{A}|$ of a quaternion $\mathbb{A}$ is computed by $\sqrt{\mathbb{A} \circ \overline{\mathbb{A}}} \equiv \sqrt{\overline{\mathbb{A}} \circ \mathbb{A}}$ (a square root of the sum of squares of its four components). Clearly, $|\mathbb{A} \circ \mathbb{B}| = \sqrt{\mathbb{A} \circ \mathbb{B} \circ \overline{\mathbb{B}} \circ \overline{\mathbb{A}}} = |\mathbb{A}|\,|\mathbb{B}|$. For a vector, this implies $f^2 = \mathbf{f} \circ \overline{\mathbf{f}} \equiv \overline{\mathbf{f}} \circ \mathbf{f}$.

For COMPLEX quantities (with no $\mathfrak{j}$ and $\mathfrak{k}$ components), we use the *calligraphic* type (e.g. $\mathcal{D}$), an asterisk implying the complex conjugation, thus: $\mathcal{D}^*$. In subsequent chapters, we make an exception to this rule by employing the symbols z and q for complex *variables.*

1.1.2 Rotations

The process of rotation in three dimensions is greatly simplified in the quaternion-algebra representation.

First we note that any quaternion $\mathbb{A}$ can be written as $A + a\,\widehat{\mathbf{a}}$ where a and $\widehat{\mathbf{a}}$ are the length and unit direction, respectively, of $\mathbf{a}$. Since $\widehat{\mathbf{a}}^2 = -1$, $\widehat{\mathbf{a}}$ behaves like a single imaginary unit; the whole quaternion thus has attributes of a complex number. Correspondingly, we can evaluate *functions* of $\mathbb{A}$, e.g.

$$e^{\mathbb{A}} = e^{A}(\cos a + \widehat{\mathbf{a}} \sin a) \tag{1.6}$$

Now, let us assume that a rotation is defined by a vector $\mathbf{w} \equiv w_z\mathfrak{i} + w_y\mathfrak{j} + w_x\mathfrak{k}$, with its unit direction $\widehat{\mathbf{w}}$ specifying the rotation's axis (passing through the origin) and its length w representing the corresponding angle.

One can easily verify that a point $\mathbf{x}$ (also represented by a pure quaternion) is moved, under such a rotation, to a new point $\mathbf{x}^\dagger$ by

$$\mathbf{x}^\dagger = e^{-\mathbf{w}/2} \circ \mathbf{x} \circ e^{\mathbf{w}/2} \equiv \overline{\mathbb{R}} \circ \mathbf{x} \circ \mathbb{R} \tag{1.7}$$

Proof. Let us rewrite $\mathbf{x}$ as $\mathbf{x}_{\|}+\mathbf{x}_{\perp}$, the parallel and perpendicular components with respect to $\widehat{\mathbf{w}}$. Obviously, $\mathbf{x}_{\|}$ commutes with $\mathbf{w}$ (in terms of quaternion multiplication), whereas $\mathbf{x}_{\perp}$ and $\mathbf{w}$ anticommute (i.e. $\mathbf{x}_{\perp} \circ \mathbf{w} = -\mathbf{w} \circ \mathbf{x}_{\perp} = \mathbf{w} \times \mathbf{x}_{\perp}$), implying that

$$\mathbf{x}_{\perp} \circ e^{\mathbf{w}/2} = e^{-\mathbf{w}/2} \circ \mathbf{x}_{\perp} \tag{1.8}$$

since $\mathbf{x}_{\perp} \circ \mathbf{w}^n = (-1)^n \mathbf{w}^n \circ \mathbf{x}_{\perp}$, for each nonnegative integer n. One thus gets

$$\mathbf{x}^\dagger = e^{-\mathbf{w}/2} \circ e^{\mathbf{w}/2} \circ \mathbf{x}_{\|} + e^{-\mathbf{w}/2} \circ e^{-\mathbf{w}/2} \circ \mathbf{x}_{\perp} = \mathbf{x}_{\|} + e^{-\mathbf{w}} \circ \mathbf{x}_{\perp} \tag{1.9}$$

Furthermore, since $e^{-\mathbf{w}} = \cos w - \widehat{\mathbf{w}} \sin w$, we finally get

$$\mathbf{x}^\dagger = \mathbf{x}_{\|} + \mathbf{x}_{\perp} \cos w + (\widehat{\mathbf{w}} \times \mathbf{x}_{\perp}) \sin w \tag{1.10}$$

which is exactly what a right-handed rotation should do ($\widehat{\mathbf{w}} \times \mathbf{x}_{\perp}$ is $\mathbf{x}_{\perp}$, rotated by 90°). □

In (1.7), $\mathbb{R}$ obviously meets

$$\overline{\mathbb{R}} \circ \mathbb{R} = \mathbb{R} \circ \overline{\mathbb{R}} = 1 \tag{1.11}$$

(i.e. $\mathbb{R}^{-1} \equiv \overline{\mathbb{R}}$), since the powers of $\mathbf{w}$ commute with one another, and the law of exponents then (and only then) implies that $e^{-\mathbf{w}} \circ e^{\mathbf{w}} = e^{\mathbf{w}-\mathbf{w}} = e^0 = 1$.

The reverse is also true; any $\mathbb{R}$ which meets (1.11) represents a specific rotation and must have the form of $\cos(w/2) + \widehat{\mathbf{w}} \sin(w/2) = e^{\mathbf{w}/2}$.

Proof. Any quaternion can be written as $A + a\widehat{\mathbf{a}}$, where, in the case of $\mathbb{R}$, we also have $A^2 + a^2 = 1$. This further implies that there is α such that $A = \cos(\alpha)$ and $a = \sin(\alpha)$. All we need to do is to identify $w/2$ with α and $\widehat{\mathbf{w}}$ with $\widehat{\mathbf{a}}$. □

Note that $-\mathbb{R}$ represents the same rotation as $\mathbb{R}$, since rotating by an angle w around a unit direction $\widehat{\mathbf{w}}$ yields the same result as rotating by $2\pi - w$ around $-\widehat{\mathbf{w}}$, and $\cos(\pi - w/2) - \widehat{\mathbf{w}}\sin(\pi - w/2)$ reduces to $-\cos(w/2) - \widehat{\mathbf{w}}\sin(w/2)$.

Also note that, when a *quaternion* quantity $\mathbb{A}$ is transformed via $\overline{\mathbb{R}} \circ \mathbb{A} \circ \mathbb{R}$, this results in the corresponding rotation of the vector part of $\mathbb{A}$, leaving its scalar (real) part *unchanged* (INVARIANT).

1.1.2.1 *Time-dependent rotations*

Using $\mathbb{R} = e^{\frac{\alpha}{2}s}$, where s denotes time, thus yields a rotation with a *uniform* ANGULAR VELOCITY α. Note that when differentiating

$$\mathbf{x}^{\dagger} = e^{-\frac{\alpha}{2}s} \circ \mathbf{x} \circ e^{\frac{\alpha}{2}s} \tag{1.12}$$

with respect to s, we get

$$\mathbf{x}^{\dagger} \circ \frac{\alpha}{2} - \frac{\alpha}{2} \circ \mathbf{x}^{\dagger} \equiv \alpha \times \mathbf{x}^{\dagger} \tag{1.13}$$

Similarly, differentiating $\mathbf{x}^{\dagger} = \overline{\mathbb{R}} \circ \mathbf{x} \circ \mathbb{R}$, where $\mathbb{R}$ is now an *arbitrary* function of s satisfying $\mathbb{R} \circ \overline{\mathbb{R}} = 1$, we get

$$\mathbf{x}^{\dagger} \circ \overline{\mathbb{R}} \circ \mathbb{R}' + \overline{\mathbb{R}}' \circ \mathbb{R} \circ \mathbf{x}^{\dagger} = \mathbf{x}^{\dagger} \circ \overline{\mathbb{R}} \circ \mathbb{R}' - \overline{\mathbb{R}} \circ \mathbb{R}' \circ \mathbf{x}^{\dagger} = (2\overline{\mathbb{R}} \circ \mathbb{R}') \times \mathbf{x}^{\dagger} \tag{1.14}$$

which follows from $(\overline{\mathbb{R}} \circ \mathbb{R})' = \overline{\mathbb{R}} \circ \mathbb{R}' + \overline{\mathbb{R}}' \circ \mathbb{R} = 0$. This clearly implies that

$$\mathbf{Z} = 2\overline{\mathbb{R}} \circ \mathbb{R}' \tag{1.15}$$

is the corresponding (instantaneous) angular velocity (it must be a vector, since $\overline{\mathbf{Z}} = -\mathbf{Z}$).

This velocity can be expressed in a special (in the context of satellite orbits, it is called KEPLER) FRAME, defined as the original frame, itself rotated by $\mathbb{R}$, thus:

$$\mathbf{Z}_{\text{o}} = \mathbb{R} \circ \mathbf{Z} \circ \overline{\mathbb{R}} = 2\mathbb{R}' \circ \overline{\mathbb{R}} \tag{1.16}$$

(the subscript 'o' will be reserved for Kepler-frame vectors), since rotating a *frame* by $\mathbb{R}$ is the same as rotating all vectors of the old frame by $\mathbb{R}^{-1} \equiv \overline{\mathbb{R}}$.

1.1.2.2 *Euler angles*

To define EULER ANGLES, we represent a general rotation $\mathbb{R}$ by three consecutive special rotations, using $\mathfrak{i}$, $\mathfrak{k}$ and $\mathfrak{i}$ again as the respective axes, thus:

$$\mathbb{R} = e^{\mathfrak{i}\frac{\psi}{2}} \circ e^{\mathfrak{k}\frac{\theta}{2}} \circ e^{\mathfrak{i}\frac{\phi}{2}} = \cos\frac{\theta}{2}\exp(\mathfrak{i}\frac{\phi+\psi}{2}) + \mathfrak{k}\sin\frac{\theta}{2}\exp(\mathfrak{i}\frac{\phi-\psi}{2}) \tag{1.17}$$

When each of the Euler angles is a function of s, we get

$$\begin{aligned}
\mathbf{Z}_{\text{o}} &= 2\left(\mathfrak{i} \circ \frac{\psi'}{2}e^{\mathfrak{i}\frac{\psi}{2}} \circ e^{\mathfrak{k}\frac{\theta}{2}} \circ e^{\mathfrak{i}\frac{\phi}{2}} + e^{\mathfrak{i}\frac{\psi}{2}} \circ \mathfrak{k}\frac{\theta'}{2}e^{\mathfrak{k}\frac{\theta}{2}} \circ e^{\mathfrak{i}\frac{\phi}{2}} + e^{\mathfrak{i}\frac{\psi}{2}} \circ e^{\mathfrak{k}\frac{\theta}{2}} \circ \mathfrak{i}\frac{\phi'}{2}e^{\mathfrak{i}\frac{\phi}{2}}\right) \circ e^{-\mathfrak{i}\frac{\phi}{2}} \circ e^{-\mathfrak{k}\frac{\theta}{2}} \circ e^{-\mathfrak{i}\frac{\psi}{2}} \\
&= \mathfrak{i}\psi' + e^{\mathfrak{i}\psi} \circ \mathfrak{k}\theta' + e^{\mathfrak{i}\frac{\psi}{2}} \circ e^{\mathfrak{k}\theta} \circ e^{-\mathfrak{i}\frac{\psi}{2}} \circ \mathfrak{i}\phi' \\
&= \mathfrak{i}(\psi' + \phi'\cos\theta) + \mathfrak{j}(-\theta'\sin\psi + \phi'\cos\psi\sin\theta) + \mathfrak{k}(\theta'\cos\psi + \phi'\sin\psi\sin\theta)
\end{aligned} \tag{1.18}$$

using the analog of (1.8). Solving for ϕ', θ' and ψ' we obtain

$$\phi' = \frac{Z_1 \sin\psi + Z_2 \cos\psi}{\sin\theta} \tag{1.19a}$$

$$\theta' = Z_1 \cos\psi - Z_2 \sin\psi \tag{1.19b}$$

$$\psi' = Z_3 - \phi' \cos\theta \tag{1.19c}$$

where Z_1, Z_2 and Z_3 are the individual components of $\mathbf{Z}_\text{o}$ (to simplify our notation, we drop the Kepler-frame subscript when referring to these).

As an interesting exercise we prove

$$\mathbb{R} \equiv e^{\mathfrak{i}\frac{\psi}{2}} \circ e^{\mathfrak{k}\frac{\theta}{2}} \circ e^{\mathfrak{i}\frac{\phi}{2}} = e^{\mathfrak{i}\frac{\phi}{2}} \circ e^{\mathbf{n}\frac{\theta}{2}} \circ e^{\boldsymbol{\sigma}\frac{\psi}{2}} \tag{1.20}$$

where

$$\mathbf{n} = e^{-\mathfrak{i}\frac{\phi}{2}} \circ \mathfrak{k} \circ e^{\mathfrak{i}\frac{\phi}{2}} \tag{1.21}$$

is the so called NODAL direction, and

$$\boldsymbol{\sigma} = \overline{\mathbb{R}} \circ \mathfrak{i} \circ \mathbb{R} = e^{-\mathbf{n}\frac{\theta}{2}} \circ \mathfrak{i} \circ e^{\mathbf{n}\frac{\theta}{2}} \tag{1.22}$$

is the new direction of $\mathfrak{i}$. Note that (1.20) is easy to interpret and understand geometrically.

Proof. Equation (1.21) implies that

$$f(\mathbf{n}) = e^{-\mathfrak{i}\frac{\phi}{2}} \circ f(\mathfrak{k}) \circ e^{\mathfrak{i}\frac{\phi}{2}} \Rightarrow f(\mathfrak{k}) \circ e^{\mathfrak{i}\frac{\phi}{2}} = e^{\mathfrak{i}\frac{\phi}{2}} \circ f(\mathbf{n}) \tag{1.23}$$

where f is an arbitrary function. This proves that

$$\overline{\mathbb{R}} \circ \mathfrak{i} \circ \mathbb{R} = e^{-\mathfrak{i}\frac{\phi}{2}} \circ e^{-\mathfrak{k}\frac{\theta}{2}} \circ e^{-\mathfrak{i}\frac{\psi}{2}} \circ \mathfrak{i} \circ e^{\mathfrak{i}\frac{\psi}{2}} \circ e^{\mathfrak{k}\frac{\theta}{2}} \circ e^{\mathfrak{i}\frac{\phi}{2}} = e^{-\mathbf{n}\frac{\theta}{2}} \circ \mathfrak{i} \circ e^{\mathbf{n}\frac{\theta}{2}} \tag{1.24}$$

Now, based on (1.22), we get

$$e^{\mathbf{n}\frac{\theta}{2}} \circ f(\boldsymbol{\sigma}) = f(\mathfrak{i}) \circ e^{\mathbf{n}\frac{\theta}{2}} \tag{1.25}$$

which, together with (1.23), establishes

$$e^{\mathfrak{i}\frac{\phi}{2}} \circ e^{\mathbf{n}\frac{\theta}{2}} \circ e^{\boldsymbol{\sigma}\frac{\psi}{2}} = e^{\mathfrak{i}\frac{\phi}{2}} \circ e^{\mathfrak{i}\frac{\psi}{2}} \circ e^{\mathbf{n}\frac{\theta}{2}} = e^{\mathfrak{i}\frac{\psi}{2}} \circ e^{\mathfrak{i}\frac{\phi}{2}} \circ e^{\mathbf{n}\frac{\theta}{2}} = e^{\mathfrak{i}\frac{\psi}{2}} \circ e^{\mathfrak{k}\frac{\theta}{2}} \circ e^{\mathfrak{i}\frac{\phi}{2}} \tag{1.26}$$

□

1.2 QUATERNIONIC FORMULATION OF KEPLER PROBLEM

Our main objective is to solve the following perturbed KEPLER PROBLEM

$$\ddot{\mathbf{r}} + \mu \frac{\mathbf{r}}{r^3} = \varepsilon \mathbf{f} \tag{1.27}$$

where $\mathbf{r}$ is the usual shorthand for a vector with three components, say $\mathbf{r}_x(t)$, $\mathbf{r}_y(t)$ and $\mathbf{r}_z(t)$, each a function of time t, $r \equiv |\mathbf{r}|$, the two dots imply double differentiation with respect to t, μ is a constant, and $\varepsilon\mathbf{f}$ is small. We will refer to $\varepsilon\mathbf{f}$ as the *perturbing force* (per unit mass of the perturbed body); $\mathbf{f}$ is usually a function of both $\mathbf{r}$ and t. Note that we can interpret both $\mathbf{r}$ and $\mathbf{f}$ as pure-vector quaternions; the equation can be thus seen as a quaternionic equation (missing a scalar part).

The precise form of various perturbing forces is discussed in the next chapter. We now concentrate on the strictly mathematical issue of solving (1.27).

1.2.1 Transformed equation

We proceed to rewrite the equation by means of quaternion algebra.

First we introduce a new *dependent* quaternion variable $\mathbb{U}$ (which has an extra fourth dimension, compared to the old $\mathbf{r}$) and a new *independent* variable s called MODIFIED TIME (dimensionless). These will be related to the old (dependent and independent) variables in the following way:

$$\mathbf{r} = \overline{\mathbb{U}} \circ \mathfrak{k} \circ \mathbb{U} \tag{1.28a}$$

$$\frac{dt}{ds} = 2r\sqrt{\frac{a}{\mu}} \equiv rH \tag{1.28b}$$

where $a > 0$ is a real, and at this point arbitrary, function of s (its eventual choice will enable us to simplify the corresponding solution). Note that (1.28a) implies that $r = |\mathbf{r}| \equiv \sqrt{-\mathbf{r} \circ \mathbf{r}} = \overline{\mathbb{U}} \circ \mathbb{U} = \mathbb{U} \circ \overline{\mathbb{U}}$.

When (1.28) are substituted into the original equation (1.27), we obtain

$$2\mathbb{U}'' - (2\mathbb{U}' \circ \overline{\mathbb{U}}' - 4a)\frac{\mathbb{U}}{r} + 2\mathfrak{k} \circ \mathbb{U}'\frac{\Gamma}{r} + \mathfrak{k} \circ \mathbb{U}\left(\frac{\Gamma}{r}\right)' - \left(\mathbb{U}' + \mathfrak{k} \circ \mathbb{U}\frac{\Gamma}{2r}\right)\frac{a'}{a} + 4\frac{a}{\mu}\varepsilon \mathbb{U} \circ \mathbf{r} \circ \mathbf{f} = 0 \tag{1.29}$$

where

$$\Gamma \equiv \overline{\mathbb{U}} \circ \mathfrak{k} \circ \mathbb{U}' - \overline{\mathbb{U}}' \circ \mathfrak{k} \circ \mathbb{U} \tag{1.30}$$

(real), and the prime indicates differentiation with respect to s. Note that Γ can be also expressed as $2\mathrm{Re}(\overline{\mathbb{U}} \circ \mathfrak{k} \circ \mathbb{U}')$, where $\mathrm{Re}(\mathbb{A}) \equiv (\mathbb{A} + \overline{\mathbb{A}})/2$ returns the real (*scalar*) part of its argument. Note that $\mathrm{Re}(\mathbb{A} \circ \mathbb{B}) = \mathrm{Re}(\mathbb{B} \circ \mathbb{A})$.

Proof.

$$\dot{\mathbf{r}} = 2\overline{\mathbb{U}} \circ \mathfrak{k} \circ \overset{\bullet}{\mathbb{U}} - \frac{\Gamma}{rH} \tag{1.31}$$

(dividing by rH converts $'$ to $^\bullet$), premultiplied by $H\,\mathfrak{k} \circ \mathbb{U}$, implies

$$H\,\mathfrak{k} \circ \mathbb{U} \circ \dot{\mathbf{r}} = -2\mathbb{U}' - \frac{\mathfrak{k} \circ \mathbb{U}}{r}\Gamma \tag{1.32}$$

Applying $rH\frac{d}{dt} \equiv \frac{d}{ds}$ to the last equation results in

$$rH\,(H\,\mathfrak{k} \circ \mathbb{U} \circ \dot{\mathbf{r}})^\bullet = -2\mathbb{U}'' - \left(\frac{\mathfrak{k} \circ \mathbb{U}}{r}\Gamma\right)' \tag{1.33}$$

Expanding the left hand side results in

$$\begin{aligned} & rH^2\,\mathfrak{k} \circ \mathbb{U} \circ \left(\varepsilon\mathbf{f} - \mu\frac{\mathbf{r}}{r^3}\right) + H'\,\mathfrak{k} \circ \mathbb{U} \circ \dot{\mathbf{r}} + H\,\mathfrak{k} \circ \mathbb{U}' \circ \dot{\mathbf{r}} \\ & = 4\frac{a}{\mu}\varepsilon\,\mathbb{U} \circ \mathbf{r} \circ \mathbf{f} + 4a\frac{\mathbb{U}}{r} + \frac{a'}{a}\left(-\mathbb{U}' - \frac{\mathfrak{k} \circ \mathbb{U}}{2r}\Gamma\right) + \frac{\mathfrak{k} \circ \mathbb{U}'}{r} \circ (2\overline{\mathbb{U}}' \circ \mathfrak{k} \circ \mathbb{U} + \Gamma) \end{aligned} \tag{1.34}$$

since

$$\mathfrak{k} \circ \mathbb{U} \equiv \frac{\mathbb{U} \circ \mathbf{r}}{r} \tag{1.35a}$$

$$\mathbf{r} \circ \mathbf{r} = -r^2 \tag{1.35b}$$

$$\frac{H'}{H} \equiv \frac{a'}{2a} \tag{1.35c}$$

and

$$\dot{\mathbf{r}} = 2\overline{\overset{\bullet}{\mathbb{U}}} \circ \mathfrak{k} \circ \mathbb{U} + \frac{\Gamma}{rH} \tag{1.36}$$

Substituting (1.34) for the left hand side of (1.33) yields (1.29). □

The major advantage of (1.29) over (1.27) is obviously *not* the relative complexity of its appearance, but the ease with which its solution can be constructed.

1.2.2 Gauge factor

Since (1.29) is four dimensional (as compared to three dimensions of Eq. 1.27), it is obvious that its solution cannot be unique and its general form must contain an arbitrary GAUGE factor. This is indeed the case, made more precise by the following statement:

Remark. If $\mathbb{U}$ solves (1.29), so does $e^{\mathfrak{k}\alpha} \circ \mathbb{U}$, where α is an *arbitrary* real function of s.

Proof. The first and second derivative of $\mathbb{U}_{\text{new}} \equiv e^{\mathfrak{k}\alpha} \circ \mathbb{U}$ is equal to

$$\mathfrak{k}\alpha' \circ e^{\mathfrak{k}\alpha} \circ \mathbb{U} + e^{\mathfrak{k}\alpha} \circ \mathbb{U}' \tag{1.37}$$

and to

$$\mathfrak{k}\,\alpha'' \circ e^{\mathfrak{k}\,\alpha} \circ \mathbb{U} - (\alpha')^2 e^{\mathfrak{k}\alpha} \circ \mathbb{U} + 2\mathfrak{k}\alpha' \circ e^{\mathfrak{k}\alpha} \circ \mathbb{U}' + e^{\mathfrak{k}\,\alpha} \circ \mathbb{U}'' \tag{1.38}$$

respectively. This implies that

$$\Gamma_{\text{new}} = \Gamma - 2\alpha' r \tag{1.39}$$

and

$$\overline{\mathbb{U}}'_{\text{new}} \circ \mathbb{U}'_{\text{new}} = \overline{\mathbb{U}}' \circ \mathbb{U}' - \alpha'\Gamma + (\alpha')^2 r \tag{1.40}$$

Substituting these back into (1.29), terms proportional to α'', α' and $(\alpha')^2$ all cancel out, namely

$$2\mathfrak{k}\,\alpha'' \circ e^{\mathfrak{k}\,\alpha} \circ \mathbb{U} + \mathfrak{k} \circ e^{\mathfrak{k}\,\alpha} \circ \mathbb{U}(-2\alpha'') = 0 \tag{1.41}$$

$$\begin{aligned} &4\mathfrak{k}\,\alpha' \circ e^{\mathfrak{k}\,\alpha} \circ \mathbb{U}' - 2\left(\overline{\mathbb{U}}' \circ e^{-\mathfrak{k}\,\alpha} \circ \mathfrak{k}\,\alpha' \circ e^{\mathfrak{k}\,\alpha} \circ \mathbb{U} - \overline{\mathbb{U}} \circ e^{-\mathfrak{k}\,\alpha} \circ \alpha'\mathfrak{k} \circ e^{\mathfrak{k}\,\alpha} \circ \mathbb{U}'\right) \circ \frac{e^{\mathfrak{k}\,\alpha} \circ \mathbb{U}}{r} \\ &-2\alpha' e^{\mathfrak{k}\,\alpha} \circ \mathbb{U}\frac{\Gamma}{r} + 2\mathfrak{k} \circ e^{\mathfrak{k}\,\alpha} \circ \mathbb{U}'(-2\alpha') - \left[\mathfrak{k}\,\alpha' \circ e^{\mathfrak{k}\,\alpha} \circ \mathbb{U} + \mathfrak{k} \circ e^{\mathfrak{k}\,\alpha} \circ \mathbb{U}\,(-\alpha')\right] \frac{a'}{a} = 0 \end{aligned} \tag{1.42}$$

and

$$-2(\alpha')^2 e^{\mathfrak{k}\,\alpha} \circ \mathbb{U} - 2\overline{\mathbb{U}} \circ e^{-\mathfrak{k}\,\alpha}(\alpha')^2 \circ e^{\mathfrak{k}\,\alpha} \circ \mathbb{U} \circ \frac{e^{\mathfrak{k}\,\alpha} \circ \mathbb{U}}{r} - 2\alpha' e^{\mathfrak{k}\,\alpha} \circ \mathbb{U}\,(-2\alpha') = 0 \tag{1.43}$$

respectively, and we are left with the original equation premultiplied by $e^{\mathfrak{k}\,\alpha}$. □

Note that both r and $\mathbf{r}$ (and consequently $\mathbf{f}$) remain invariant under the $\mathbb{U} \to e^{\mathfrak{k}\alpha}\mathbb{U}$ transformation. This makes α a physically inconsequential parameter. We can adjust it to give our solution a particularly simple form (as done in subsequent sections).

In this context, we should also mention that for numerical integration of (1.29) — not a topic of this book — a particularly simple constraint (called GAUGE) regarding the choice of α is usually employed, namely $\Gamma \equiv 0$ (that this is always possible should be clear from Eq. 1.39). Such a choice dramatically simplifies

(1.29), furthermore, the new equation ensures that the $\Gamma \equiv 0$ condition is maintained *automatically* (as long as it is met *initially*), since Γ becomes a constant of motion.

Proof.

$$\Gamma' = \overline{\mathbb{U}} \circ \mathfrak{k} \circ \mathbb{U}'' - \overline{\mathbb{U}}'' \circ \mathfrak{k} \circ \mathbb{U} = 2\mathrm{Re}\left(\overline{\mathbb{U}} \circ \mathfrak{k} \circ \mathbb{U}''\right) \tag{1.44}$$

Substituting $\Gamma = 0$ into (1.29) results in

$$2\mathbb{U}'' = (2\mathbb{U}' \circ \overline{\mathbb{U}}' - 4a)\frac{\mathbb{U}}{r} + \mathbb{U}'\frac{a'}{a} - 4\frac{a}{\mu}\varepsilon\mathbb{U} \circ \mathbf{r} \circ \mathbf{f} \tag{1.45}$$

When premultiplied by $\overline{\mathbb{U}} \circ \mathfrak{k}$, the first and last term of the right hand side clearly yield a vector (with a zero real part). The second term, after its real part is taken, becomes $\Gamma a'/a$, which is also equal to zero. □

Clearly, the easiest choice of a is to make it equal to $\mu/4$ (a constant), which simplifies (1.45) even more, thus:

$$2\mathbb{U}'' = E_{\mathrm{K}}\mathbb{U} - \varepsilon\mathbb{U} \circ \mathbf{r} \circ \mathbf{f} \tag{1.46}$$

where

$$E_{\mathrm{K}} \equiv \frac{2\mathbb{U}' \circ \overline{\mathbb{U}}' - \mu}{r} = \frac{\dot{\mathbf{r}}^2}{2} - \frac{\mu}{r} \tag{1.47}$$

is the so-called Kepler energy . This equation is easy to integrate numerically, and there is yet another extra benefit: the *actual* value of Γ (which, sooner or later becomes non-zero, due to round off errors) can be used as an excellent measure of the solution's accuracy.

However, using the same approach to build an analytic solution would prove too restrictive (to the point of making it practically impossible). So, for our purpose, the very attractive $\Gamma \equiv 0$ gauge is a dead end, and has to be replaced by a different constraint.

1.3 SOLVING UNPERTURBED KEPLER PROBLEM

As a preliminary to our main task of solving the perturbed equation (1.29), we first discuss the solution to the UNPERTURBED case.[2] The result will be needed as the first step to building a perturbed solution.

1.3.1 Quaternionic solution

When the perturbing force $\mathbf{f}$ is identically equal to zero, we may choose a to be constant and $\Gamma \equiv 0$. Equation (2.6) then reads

$$\mathbb{U}'' - (\mathbb{U}' \circ \overline{\mathbb{U}}' - 2a)\frac{\mathbb{U}}{r} = 0 \tag{1.48}$$

It easily follows that

$$\frac{\mathbb{U}' \circ \overline{\mathbb{U}}' - 2a}{r} = \frac{2a}{\mu}\left(\frac{v^2}{2} - \frac{\mu}{r}\right) \tag{1.49}$$

is a constant of motion, since its derivative

$$\frac{\mathbb{U}'' \circ \overline{\mathbb{U}}' + \mathbb{U}' \circ \overline{\mathbb{U}}''}{r} - \frac{\mathbb{U}' \circ \overline{\mathbb{U}}' - 2a}{r^2}(\mathbb{U} \circ \overline{\mathbb{U}}' + \mathbb{U}' \circ \overline{\mathbb{U}}) \tag{1.50}$$

[2]See the following references: [4], [8], [12], [18], [27], [30], [31], [33], [34] and [51].

is clearly equal to zero; to see that, substitute $\mathbb{U}''$ from (1.48). Here, we are interested only in situations where this constant is negative (the satellite is *orbiting* the primary). Furthermore, we can always choose a (based on the initial conditions) to make this constant equal to -1. The equation to solve is thus

$$\mathbb{U}'' + \mathbb{U} = 0 \tag{1.51}$$

Its general solution is clearly $\mathbb{U} = \widetilde{\mathbb{P}} \sin s + \widetilde{\mathbb{Q}} \cos s$ or, equivalently,

$$\mathbb{U} = e^{\mathrm{i}s} \circ \mathbb{P} + e^{-\mathrm{i}s} \circ \mathbb{Q} \tag{1.52}$$

where $\widetilde{\mathbb{P}}$ and $\widetilde{\mathbb{Q}}$, and $\mathbb{P}$ and $\mathbb{Q}$, are constant quaternions, subject only to the $\Gamma = 0$ constraint, and the

$$\frac{\mathbb{U}' \circ \overline{\mathbb{U}}' - 2a}{r} = -1 \tag{1.53}$$

condition. We will show in the next section that $e^{\mathrm{i}s} \circ \mathbb{P} + e^{-\mathrm{i}s} \circ \mathbb{Q}$ can be parametrized by

$$e^{\mathfrak{k}\alpha} \circ \left(A(1+\gamma \mathrm{j}) \circ e^{\mathrm{i}(s-s_{\mathrm{p}})} + B\, e^{-\mathrm{i}(s-s_{\mathrm{p}})} \right) \circ \mathbb{R} \tag{1.54}$$

where α, A, B, s_{p} and γ are *real* constants, and $\mathbb{R}$ is a constant quaternion such that $\mathbb{R} \circ \overline{\mathbb{R}} = 1$. It is easy to verify that this solution results in $\Gamma = 4\gamma A^2$, meeting our original assumption only when $\gamma = 0$. After discarding the physically irrelevant $e^{\mathfrak{k}\alpha}$ factor (by setting $\alpha = 0$), our final solution reads

$$\mathbb{U} = \left(A\, e^{\mathrm{i}(s-s_{\mathrm{p}})} + B\, e^{-\mathrm{i}(s-s_{\mathrm{p}})} \right) \circ \mathbb{R} \equiv \mathcal{U}_0 \circ \mathbb{R} \tag{1.55}$$

where we still have to make sure that $(\mathbb{U}' \circ \overline{\mathbb{U}}' - 2a)/r = -1$ or, equivalently, $\mathbb{U}' \circ \overline{\mathbb{U}}' + \mathbb{U} \circ \overline{\mathbb{U}} = 2a$. Since

$$\mathbb{U}' \circ \overline{\mathbb{U}}' = A^2 + B^2 - A\ B\ e^{2\mathrm{i}(s-s_{\mathrm{p}})} - A\ B\ e^{-2\mathrm{i}(s-s_{\mathrm{p}})} \tag{1.56}$$

and

$$\mathbb{U} \circ \overline{\mathbb{U}} = A^2 + B^2 + A\ B\ e^{2\mathrm{i}(s-s_{\mathrm{p}})} + A\ B\ e^{-2\mathrm{i}(s-s_{\mathrm{p}})} \tag{1.57}$$

this implies that $a \equiv A^2 + B^2$. We prove shortly that a represents the SEMIMAJOR AXIS of the resulting *ellipse*, $2AB/(A^2 + B^2)$ is its ECCENTRICITY, s_{p} is the value of s at APOCENTER (the point of greatest distance from the origin), and $\mathbb{R}$ defines the ellipse's ATTITUDE, usually parametrized by the three Euler angles of (1.17).

From now on, we find it convenient to use a and $\beta \equiv B/A$ (MODIFIED ECCENTRICITY) as the basic parameters of the solution (instead of the original A and B), thus:

$$\mathbb{U} = \sqrt{\frac{a}{1+\beta^2}} \left[e^{\mathrm{i}(s-s_{\mathrm{p}})} + \beta\ e^{-\mathrm{i}(s-s_{\mathrm{p}})} \right] \circ \mathbb{R} \tag{1.58}$$

Up to the physically redundant factor of $e^{\mathfrak{k}\alpha}$, this represents the most general solution to the unperturbed equation.

Expressing $\mathbf{r}$ and r in terms of the new parameters results in

$$\begin{aligned} \mathbf{r} &= \overline{\mathbb{R}} \circ \mathfrak{k} \circ \frac{a}{1+\beta^2} z \left(1 + \frac{\beta}{z}\right)^2 \circ \mathbb{R} = \overline{\mathbb{R}} \circ \mathfrak{k} \circ \frac{a}{1+\beta^2} \left(z + \frac{\beta^2}{z} + 2\beta \right) \circ \mathbb{R} \\ &= \overline{\mathbb{R}} \circ a \left[\mathfrak{k} \left(\cos[2(s-s_{\mathrm{p}})] + \frac{2\beta}{1+\beta^2} \right) + \mathrm{j} \frac{1-\beta^2}{1+\beta^2} \sin[2(s-s_{\mathrm{p}})] \right] \circ \mathbb{R} \end{aligned} \tag{1.59}$$

where $z \equiv e^{2\mathrm{i}(s-s_{\mathrm{p}})}$, and

$$r = \frac{a}{1+\beta^2}(1+\beta z)\left(1 + \frac{\beta}{z}\right) = a\left[1 + \frac{\beta}{1+\beta^2}\left(z + \frac{1}{z}\right)\right] = a\left(1 + \frac{2\beta}{1+\beta^2}\cos[2(s-s_{\mathrm{p}})]\right) \tag{1.60}$$

From now on, we always identify the parameter a (semi-major axis) of our solution with the 'arbitrary' a in (1.28b) and (1.29).

This is all we need to tackle the perturbed equation. Readers who are not interested in classical details of the unperturbed solution may proceed directly to the last section of this chapter.

1.3.1.1 Kepler equation

So far, we have presented our solution as a function of modified time s. Regular time t can be brought back by integrating (1.28b), which yields

$$t = 2\sqrt{\frac{a^3}{\mu}}\left(s - s_{\mathrm{p}} + \frac{\beta}{1+\beta^2}\sin[2(s-s_{\mathrm{p}})]\right) + t_0 = \sqrt{\frac{a^3}{\mu}}\,(\omega + e\sin\omega) + t_0 \tag{1.61}$$

where e is the orbit's eccentricity, $\omega \equiv 2(s - s_{\mathrm{p}})$, and t_0 is the value of *regular* time at apocenter.

Our $2(s - s_{\mathrm{p}})$ is thus equivalent to the so called ECCENTRIC ANOMALY, of the traditional solution (except that the former is set to zero at the orbit's *apocenter*, the latter at its *pericenter* - this is why our equations appear to have a wrong sign of e).

Using e as an iteration parameter, one can easily solve (1.61), or the equivalent

$$\omega + e\sin\omega = \tau \equiv \sqrt{\frac{\mu}{a^3}}(t - t_0) = 2\pi\frac{t-t_0}{T} \tag{1.62}$$

for ω, where τ is the so called MEAN ANOMALY, again, measured from apocenter, and T is the orbital period (note how this implies the third Kepler law $T = 2\pi\sqrt{a^3/\mu}$). This results in an infinite power series with the first few terms given by

$$\tau - e\sin\tau + \frac{e^2}{2}\sin 2\tau + \frac{e^3}{8}(\sin\tau - 3\sin 3\tau) - \frac{e^4}{6}(\sin 2\tau - 2\sin 4\tau) + \cdots \tag{1.63}$$

Substituting into (1.59), this yields (up to e^2 accuracy)

$$\mathbf{r} \simeq a\overline{\mathbb{R}} \circ \mathfrak{k}\left[\exp(\mathrm{i}\tau) + \frac{e}{2}\left(3 - \exp(2\mathrm{i}\tau)\right) + \frac{e^2}{8}\left(\exp(-\mathrm{i}\tau) - 4\exp(\mathrm{i}\tau) + 3\exp(3\mathrm{i}\tau)\right) + \cdots\right] \circ \mathbb{R} \tag{1.64}$$

(the traditional solution is recovered by $e \to -e$).

1.3.1.2 Elliptical orbit

Note that in terms of e and ω, (1.59) can be rewritten as

$$\mathbf{r} = \overline{\mathbb{R}} \circ a\left(\mathfrak{k}\,(\cos\omega + e) + \mathrm{j}\sqrt{1-e^2}\sin\omega\right) \circ \mathbb{R} \tag{1.65}$$

This clearly shows that, in Kepler frame,

$$(x - a\ e)^2 + \frac{y^2}{1-e^2} = a^2 \tag{1.66}$$

(x and y are the first two components of $\mathbf{r}_{\mathrm{o}}$), which is an equation of an ellipse with eccentricity e and semimajor axis a (first Kepler law).

Incidentally, the second Kepler law follows from

$$\frac{d(|\mathbf{r} \times \dot{\mathbf{r}}|)}{dt} = \frac{\mathbf{r} \times \dot{\mathbf{r}}}{|\mathbf{r} \times \dot{\mathbf{r}}|} \cdot \frac{d(\mathbf{r} \times \dot{\mathbf{r}})}{dt} \tag{1.67}$$

and

$$\frac{d(\mathbf{r} \times \dot{\mathbf{r}})}{dt} = (\dot{\mathbf{r}} \times \dot{\mathbf{r}}) - (\mathbf{r} \times \ddot{\mathbf{r}}) = \mu\frac{\mathbf{r} \times \mathbf{r}}{r^3} = \mathbf{0} \tag{1.68}$$

1.3.1.3 True anomaly

We should also mention that the Kepler-frame longitude of the satellite (its so called TRUE ANOMALY) is the polar-coordinate angle of $\mathbf{r}_o = \mathbb{R} \circ \mathbf{r} \circ \overline{\mathbb{R}}$, namely:

$$\chi = \arctan(\cos\omega + e, \sqrt{1-e^2}\sin\omega) \tag{1.69}$$

where the two-argument $\arctan(x, y)$ function returns $\arctan(y/x)$, taking into consideration the sign of each x and y; the computed values will thus range over the full $(-\pi, \pi]$ interval, instead of the usual $(-\pi/2, \pi/2]$. Note that (1.69) implies that

$$\mathbf{r} = \overline{\mathbb{R}} \circ r\,(\mathfrak{k}\cos\chi + \mathrm{j}\sin\chi) \circ \mathbb{R} \tag{1.70}$$

For small e, the right hand side of (1.69) expands to

$$\chi \simeq \omega - e\sin\omega + \frac{e^2}{4}\sin 2\omega + \cdots \tag{1.71}$$

in terms of the eccentric anomaly and, after substituting the first three terms of (1.63)

$$\chi \simeq \tau - 2e\sin\tau + \frac{5\,e^2}{4}\sin 2\tau + \cdots \tag{1.72}$$

in terms of the mean anomaly.

1.3.1.4 Inertial-frame location

Substituting (1.17) into (1.59) yields

$$\mathbf{r} = \frac{a}{1+\beta^2}\left(\mathrm{Im}[e^{\mathrm{i}\psi}z(1+\tfrac{\beta}{z})^2](\mathrm{i}\sin\theta + \mathrm{j}\circ e^{\mathrm{i}\phi}\cos\theta) + \mathrm{Re}[e^{\mathrm{i}\psi}z(1+\tfrac{\beta}{z})^2]\mathfrak{k}\circ e^{\mathrm{i}\phi}\right) \tag{1.73}$$

When both the eccentricity and inclination (the Euler angle θ, between the plane of the orbit and the x-y plane of our system of coordinates - this is usually the ecliptic) are small, one can expand the last expression in both β and θ, to find

$$\mathbf{r} \simeq a\,\mathrm{i}\,\mathrm{Im}[e^{\mathrm{i}\psi}(z+2\beta)]\theta + a\,\mathfrak{k}\circ e^{\mathrm{i}(\phi+\psi)}\left[z(1-\beta^2-\tfrac{\theta^2}{4}) + \tfrac{\beta^2}{z} + 2\beta\right] + a\,\mathfrak{k}\circ e^{\mathrm{i}(\phi-\psi)}\tfrac{\theta^2}{4z} + \cdots \tag{1.74}$$

Note that the first term is the location's z coordinate, and the last two terms (after discarding $\mathfrak{k}$) yield a *complex* quantity which represents both the x and y coordinates. Thus, the *inertial-frame* LONGITUDE can be computed by taking the purely-imaginary part of the (complex) logarithm of the $\mathfrak{k}$-proportional terms in (1.74), namely:

$$\begin{aligned}
&\omega + \phi + \psi + \mathrm{Im}\left[\mathrm{Log}\left(1 - \beta^2 - \tfrac{\theta^2}{4} + \tfrac{\beta^2}{z^2} + 2\tfrac{\beta}{z} + e^{-2\mathrm{i}\psi}\tfrac{\theta^2}{4z^2} + \cdots\right)\right] \\
&\simeq \omega + \phi + \psi - 2\beta\sin\omega + \beta^2\sin 2\omega - \tfrac{\theta^2}{4}\sin 2(\omega+\psi) + \cdots \\
&\simeq \tau + \phi + \psi - 4\beta\sin\tau + 5\beta^2\sin 2\tau - \tfrac{\theta^2}{4}\sin 2(\tau+\psi) + \cdots
\end{aligned} \tag{1.75}$$

after the

$$\omega \to \tau - 2\beta\sin\tau + 2\beta^2\sin 2\tau + \cdots \tag{1.76}$$

replacement. The two β terms are called, rather archaically, EQUATION OF THE CENTER, and the last term ('correcting' for a small non-zero inclination) is the REDUCTION TO THE ECLIPTIC.

Similarly, the inertial-frame LATITUDE is computed from $\arcsin(\mathbf{r}_z/r)$. To a sufficient accuracy, this yields

$$\begin{aligned}\arcsin\frac{\operatorname{Im}[e^{i\psi}(z+2\beta)]\,\theta}{1+\beta\left(z+\frac{1}{z}\right)} &\simeq \theta\sin(\omega+\psi) - \beta\,\theta\sin(2\omega+\psi) + \beta\,\theta\sin\psi \\ &\simeq \theta\sin(\tau+\psi) - 2\beta\,\theta\sin(2\tau+\psi) + 2\beta\,\theta\sin\psi \\ &= \theta\sin(\tau+\psi) - 4\beta\,\theta\cos(\tau+\psi)\sin\tau \qquad (1.77)\end{aligned}$$

1.3.2 Orbit determination

To complete our solution, we have to establish a way of finding the ORBITAL ELEMENTS a, β, t_0, ϕ, θ and ψ, based on a set of observed values of $\mathbf{r}$ and $\dot{\mathbf{r}}$ at a specific time t (the usual INITIAL-VALUE problem), or two sets of $\mathbf{r}$ observations at t_1 and t_2 (a BOUNDARY-VALUE problem). One can consider other practically important possibilities (in some cases only angular information may be available, in other cases we may have only distance-related data), but our two examples will be sufficient to demonstrate the issue.

1.3.2.1 Initial-value problem

Having established, at some time t, the value of

$$\mathbf{r} = \overline{\mathbb{R}} \circ \mathfrak{k} \circ \frac{a}{1+\beta^2}\left(z + \frac{\beta^2}{z} + 2\beta\right) \circ \mathbb{R} \qquad (1.78)$$

(Eq. 1.59) and

$$\dot{\mathbf{r}} = \frac{\overline{\mathbb{R}} \circ \mathrm{j} \circ \frac{2a}{1+\beta^2}\left(z - \frac{\beta^2}{z}\right) \circ \mathbb{R}}{2r\sqrt{\frac{a}{\mu}}} \qquad (1.79)$$

implies that we also know

$$r \equiv |\mathbf{r}| = a\left(1 + \frac{2\beta}{1+\beta^2}\cos\omega\right) \qquad (1.80)$$

(Eq. 1.60) and

$$\dot{r} = \frac{\dot{\mathbf{r}}\cdot\mathbf{r}}{r} = \frac{-\frac{4a\beta}{1+\beta^2}\sin\omega}{2r\sqrt{\frac{a}{\mu}}} \qquad (1.81)$$

(the first part follows from differentiating $r^2 \equiv \mathbf{r}\cdot\mathbf{r}$). We assume that the value of μ is known, otherwise, one additional independent observation would have to be made (and the resulting equations would have to be solved numerically).

We should mention that, in practice, one normally observes r and $\dot{r}$ together with the satellite's longitude Φ, latitude Θ and their time derivatives, and then converts this data back to $\mathbf{r}$ and $\dot{\mathbf{r}}$. The details of such a conversion are quite simple and don't need to be discussed here.

From (1.79), we get

$$|\dot{\mathbf{r}}|^2 = \left(\frac{\frac{a}{1+\beta^2}}{r\sqrt{\frac{a}{\mu}}}\right)^2\left(z - \frac{\beta^2}{z}\right)\left(\frac{1}{z} - z\beta^2\right) = \frac{a\mu}{r^2(1+\beta^2)^2}(1 - 2\beta^2\cos 2\omega + \beta^4) \qquad (1.82)$$

Now, we can verify that $|\dot{\mathbf{r}}|^2/2 - \mu/r$ (the satellite's energy, per unit mass) is indeed a constant, equal to

$$\begin{aligned}&\frac{\mu}{2a(1+\beta^2+2\beta\cos\omega)^2}\left[1-2\beta^2\cos 2\omega+\beta^4-2(1+\beta^2)(1+\beta^2+2\beta\cos\omega)\right]\\&=\frac{\mu}{2a(1+\beta^2+2\beta\cos\omega)^2}\left[-\left(1+\beta^2+2\beta\cos\omega\right)^2\right]=-\frac{\mu}{2a}\end{aligned}\tag{1.83}$$

We have thus established the value of

$$a=\left(\frac{2}{r}-\frac{|\dot{\mathbf{r}}|^2}{\mu}\right)^{-1}\tag{1.84}$$

Knowing a enables us to write

$$e\cos\omega=\frac{r}{a}-1\tag{1.85a}$$

$$e\sin\omega=-\frac{r\,\dot{r}}{\sqrt{a\mu}}\tag{1.85b}$$

based on (1.80) and (1.81), respectively. The last two equations can be easily solved for e and ω:

$$e=\sqrt{\left(\frac{r}{a}-1\right)^2+\frac{(r\dot{r})^2}{a\mu}}\tag{1.86a}$$

$$\omega=\arctan\left(\frac{r}{a}-1,-\frac{r\dot{r}}{\sqrt{a\mu}}\right)\tag{1.86b}$$

Once we have the value of e, we can convert it to

$$\beta=\frac{1-\sqrt{1-e^2}}{e}\tag{1.87}$$

Similarly, based on (1.62),

$$t_0=t-\sqrt{\frac{a^3}{\mu}}(\omega+e\sin\omega)\tag{1.88}$$

where t is the time of observation (of $\mathbf{r}$ and $\dot{\mathbf{r}}$).

Now, we still need to find the parameters (Euler angles) of the orbit's attitude $\mathbb{R}$.

Since

$$\mathbf{r}\times\dot{\mathbf{r}}=\overline{\mathbb{R}}\circ\mathfrak{i}\frac{a^2\left[1-\beta^4+2\beta(1-\beta^2)\cos\omega\right]}{(1+\beta^2)^2 r\sqrt{\frac{a}{\mu}}}\circ\mathbb{R}\tag{1.89}$$

$(\mathbf{r}\times\dot{\mathbf{r}})/|\mathbf{r}\times\dot{\mathbf{r}}|$ is clearly equal to $\overline{\mathbb{R}}\circ i\circ\mathbb{R}$ (dividing both sides of the equation by the respective magnitudes). The last expression equals

$$e^{-\mathfrak{i}\frac{\phi}{2}}\circ e^{-\mathfrak{k}\frac{\theta}{2}}\circ\mathfrak{i}\circ e^{\mathfrak{k}\frac{\theta}{2}}\circ e^{\mathfrak{i}\frac{\phi}{2}}=e^{-\mathfrak{i}\frac{\phi}{2}}\circ(\mathfrak{i}\cos\theta-\mathfrak{j}\sin\theta)\circ e^{\mathfrak{i}\frac{\phi}{2}}=\mathfrak{i}\cos\theta+\sin\theta\,(\mathfrak{k}\sin\phi-\mathfrak{j}\cos\phi)\tag{1.90}$$

Thus, we have

$$\theta=\arccos\frac{(\mathbf{r}\times\dot{\mathbf{r}})_z}{|\mathbf{r}\times\dot{\mathbf{r}}|}\tag{1.91a}$$

$$\phi=\arctan\left[-(\mathbf{r}\times\dot{\mathbf{r}})_y,(\mathbf{r}\times\dot{\mathbf{r}})_x\right]\tag{1.91b}$$

From (1.21) we get

$$\mathbf{n} = \overline{\mathbb{R}} \circ (\mathfrak{k} \cos \psi - \mathfrak{j} \sin \psi) \circ \mathbb{R} \tag{1.92}$$

Together with (1.70) this implies that $\mathbf{n} \cdot \mathbf{r} = r \cos(\psi + \chi)$.

Similarly, from (1.22) and (1.21), we get

$$\sigma \times \mathbf{n} = \overline{\mathbb{R}} \circ (\mathfrak{k} \sin \psi + \mathfrak{j} \cos \psi) \circ \mathbb{R} \tag{1.93}$$

implying that $(\sigma \times \mathbf{n}) \cdot \mathbf{r} = r \sin(\psi + \chi)$. Therefore

$$\psi + \chi = \arctan[\mathbf{n} \cdot \mathbf{r}, (\sigma \times \mathbf{n}) \cdot \mathbf{r}] \tag{1.94}$$

where

$$\chi = \arctan(\mathbf{r}_x, \mathbf{r}_y) \tag{1.95}$$

The value of all six orbital elements has thus been established.

1.3.2.2 Boundary-value problem

The observational data now consists of the satellite's two locations

$$\mathbf{r}_1 = \overline{\mathbb{R}} \circ \mathfrak{k} \circ \frac{a}{1+\beta^2}\left(z_1 + \frac{\beta^2}{z_1} + 2\beta\right) \circ \mathbb{R} \tag{1.96a}$$

$$\mathbf{r}_2 = \overline{\mathbb{R}} \circ \mathfrak{k} \circ \frac{a}{1+\beta^2}\left(z_2 + \frac{\beta^2}{z_2} + 2\beta\right) \circ \mathbb{R} \tag{1.96b}$$

at time t_1 and t_2 respectively.

This implies that

$$r_1 = a(1 + e \cos \omega_1) \tag{1.97a}$$

$$r_2 = a(1 + e \cos \omega_2) \tag{1.97b}$$

$$\mathbf{r}_1 \cdot \mathbf{r}_2 = a^2(\cos \omega_1 + e)(\cos \omega_2 + e) + a^2(1 - e^2) \sin \omega_1 \sin \omega_2 \tag{1.97c}$$

(see Eqs. 1.59 and 1.60). From (1.62) we also know that

$$t_2 - t_1 = \sqrt{\frac{a^3}{\mu}}(\omega_2 - \omega_1) + \sqrt{\frac{a^3}{\mu}} e(\sin \omega_2 - \sin \omega_1) \tag{1.98}$$

The last four equations (for four unknowns a, e, ω_1 and ω_2) can be solved only numerically. The usual procedure is to utilize LAMBERT'S THEOREM which, in a very ingenious way, reduces the set into a single (transcendental) equation for a. The details of the corresponding proof and subsequent solution are fairly elaborate (furthermore, many modern books present a very confusing picture of it); we thus attempt an alternate approach.

Let us start by reorganizing the four equations as follows:

$$\frac{r_1 + r_2}{2} = a + a\, e \cos \frac{\omega_2 - \omega_1}{2} \cos \frac{\omega_1 + \omega_2}{2} \tag{1.99a}$$

$$r_1 r_2 - \mathbf{r}_1 \cdot \mathbf{r}_2 = a^2(1 - e^2) \cos(\omega_2 - \omega_1) \tag{1.99b}$$

$$\frac{r_2 - r_1}{2a} = -e \sin \frac{\omega_2 - \omega_1}{2} \sin \frac{\omega_1 + \omega_2}{2} \tag{1.99c}$$

$$\frac{t_2 - t_1}{2}\sqrt{\frac{\mu}{a^3}} - \frac{\omega_2 - \omega_1}{2} = e \sin \frac{\omega_2 - \omega_1}{2} \cos \frac{\omega_1 + \omega_2}{2} \tag{1.99d}$$

Then we set e equal to zero, and compute, in an iterative manner

$$a = \frac{r_1 + r_2}{2} - a\,e\cos\frac{\omega_2 - \omega_1}{2}\cos\frac{\omega_1 + \omega_2}{2} \tag{1.100a}$$

$$\omega_2 - \omega_1 = 2k\pi \pm \arccos\frac{r_1 r_2 - \mathbf{r}_1 \cdot \mathbf{r}_2}{a^2(1 - e^2)} \tag{1.100b}$$

$$e = \frac{\sqrt{\left(\frac{r_2 - r_1}{2a}\right)^2 + \left(\frac{t_2 - t_1}{2}\sqrt{\frac{\mu}{a^3}} - \frac{\omega_2 - \omega_1}{2}\right)^2}}{\left|\sin\frac{\omega_2 - \omega_1}{2}\right|} \tag{1.100c}$$

$$\omega_1 + \omega_2 = 2\arctan\left(\frac{\frac{t_2 - t_1}{2}\sqrt{\frac{\mu}{a^3}} - \frac{\omega_2 - \omega_1}{2}}{\sin\frac{\omega_2 - \omega_1}{2}}, -\frac{\frac{r_2 - r_1}{2a}}{\sin\frac{\omega_2 - \omega_1}{2}}\right) \tag{1.100d}$$

where k and the sign of $\arccos(\cdots)$ in the second line have to be chosen to meet (1.98). The procedure converges relatively slowly (the error is roughly proportional to β^n, where n is the number of iterations; when $\beta \simeq 0.1$, each iterations improves the accuracy by one digit), but it is easy to understand and convert to computer code. It would be also quite simple to modify the procedure to make it quadratically convergent.

Having found a, e, ω_1 and ω_2, we can compute β and t_0 using (1.87) and (1.88) respectively (the last formula must yield the same answer with either t_1 and ω_1, or t_2 and ω_2 – one can use this as a check of the solution's correctness).

Since $\mathbf{r}_1 \times \mathbf{r}_2 = \sqrt{1 - e^2}\ \overline{\mathbb{R}} \circ \mathbf{i}\ [\sin(\omega_2 - \omega_1) + 2e(\sin\omega_2 - \sin\omega_1)] \circ \mathbb{R}$, we can now find the Euler angles by a procedure analogous to the initial-value case, i.e.

$$\theta = \arccos\frac{(\mathbf{r}_1 \times \mathbf{r}_2)_z}{|\mathbf{r}_1 \times \mathbf{r}_2|} \tag{1.101a}$$

$$\phi = \arctan\left[-(\mathbf{r}_1 \times \mathbf{r}_2)_y, (\mathbf{r}_1 \times \mathbf{r}_2)_x\right] \tag{1.101b}$$

when $\sin(\omega_2 - \omega_1) + 2e(\sin\omega_2 - \sin\omega_1)$ is positive (otherwise, we use $\mathbf{r}_2 \times \mathbf{r}_1$ instead of $\mathbf{r}_1 \times \mathbf{r}_2$) and

$$\psi = \arctan(\tilde{\mathbf{r}}_x, \tilde{\mathbf{r}}_y) \tag{1.102}$$

with

$$\tilde{\mathbf{r}} \equiv e^{\mathfrak{k}\frac{\theta}{2}} \circ e^{\mathrm{i}\frac{\phi}{2}} \circ \mathbf{r}_1 \circ e^{-\mathrm{i}\frac{\phi}{2}} \circ e^{-\mathfrak{k}\frac{\theta}{2}} \circ \left(\frac{1}{z_1} + z_1\beta^2 + 2\beta\right) \tag{1.103}$$

or, equivalently, with

$$\tilde{\mathbf{r}} \equiv e^{\mathfrak{k}\frac{\theta}{2}} \circ e^{\mathrm{i}\frac{\phi}{2}} \circ \mathbf{r}_2 \circ e^{-\mathrm{i}\frac{\phi}{2}} \circ e^{-\mathfrak{k}\frac{\theta}{2}} \circ \left(\frac{1}{z_2} + z_2\beta^2 + 2\beta\right) \tag{1.104}$$

(the two results must, again, yield the same answer).

1.4 PERTURBED-PROBLEM PRELIMINARIES

We will now return to the *perturbed* Kepler problem .

When $\varepsilon\mathbf{f}$ in (1.29) is non-zero, we extend the square brackets in (1.58) by including *all* (positive and negative) remaining *odd* powers of $q \equiv e^{\mathrm{i}(s - s_{\mathrm{p}})}$, each premultiplied (or, equivalently, postmultiplied) by a quaternionic coefficient. This results in making the corresponding $\mathbf{r}$ into a similar sum of all *even* powers of q (integer powers of $z = q^2$) with *vector* coefficients.

1.4.1 Trial solution

More explicitly, $\mathbb{U}$ will now be written as

$$\sqrt{\frac{a}{1+\beta^2}}\, q \circ \left(1 + \frac{\beta}{z} + \mathcal{D}(z) + \mathfrak{k} \circ \frac{\mathcal{S}(z)}{1+\beta\, z}\right) \circ \mathbb{R} \tag{1.105}$$

where $\mathcal{D}(z)$ and $\mathcal{S}(z)$ represent two sums of *even* powers of q with *complex* coefficients, namely

$$\mathcal{D}(z) = \cdots + \frac{\mathcal{D}_{-2}}{z^2} + \frac{\mathcal{D}_{-1}}{z} + \mathcal{D}_0 + \mathcal{D}_1 z + \mathcal{D}_2 z^2 + \cdots \tag{1.106}$$

and

$$\mathcal{S}(z) = \cdots + \frac{\mathcal{S}_{-2}}{z^2} + \frac{\mathcal{S}_{-1}}{z} + \mathcal{S}_0 + \mathcal{S}_1 z + \mathcal{S}_2 z^2 + \cdots \tag{1.107}$$

Note that $\mathcal{S}(z)$ has been divided by $1+\beta z$ for future convenience.

Later on we demonstrate that, in each step of the iteration process, the *real part* of $\mathcal{S}(z)$ can be chosen arbitrarily (the choice has no physical consequences, leaving $\mathbf{r}$ unchanged). We can thus set it equal to zero by imposing the following condition:

$$\mathcal{S}^*(z) = -\mathcal{S}(z) \tag{1.108}$$

(equivalent to $S_i = -S^*_{-i}$ for each i). This is yet another manifestation (and consequent fixing) of the $e^{\mathfrak{k}\alpha(s)}$ gauge.

Similarly, we can eliminate the coefficients $\mathcal{D}_{-1}$, $\mathcal{D}_0$, $\mathcal{S}_{-1}$ and, consequently, $\mathcal{S}_1$ (but not S_0), since we would only duplicate the

$$\sqrt{\frac{a}{1+\beta^2}}\, q \circ \left(1 + \frac{\beta}{z} + \cdots\right) \circ \mathbb{R} = q \circ \left(A + \frac{B}{z} + \cdots\right) \circ \mathbb{R} \tag{1.109}$$

part of the trial solution (note that, for convenience, we have reverted back to the old parametrization). This requires showing that (1.109) already represents $\exp(\mathfrak{i}s)$ and $\exp(-\mathfrak{i}s)$, each pre-multiplied by an *arbitrary* quaternion. More specifically, we now prove that

$$\begin{aligned} & e^{\mathfrak{k}\alpha} \circ \left[A\,(1+\mathfrak{j}\,\gamma) \circ e^{\mathfrak{i}(s-s_\mathrm{p})} + B\; e^{-\mathfrak{i}(s-s_\mathrm{p})}\right] \circ \mathbb{R} \\ &= e^{\mathfrak{i}(s-s_\mathrm{p})} \circ (A\cos\alpha - \mathfrak{i}\; A\,\gamma \sin\alpha + \mathfrak{k}\; B\sin\alpha) \circ \mathbb{R} + e^{-\mathfrak{i}(s-s_\mathrm{p})} \circ (B\cos\alpha + \mathfrak{j}\; A\,\gamma\cos\alpha + \mathfrak{k}\; A\sin\alpha) \circ \mathbb{R} \\ &= e^{\mathfrak{i}s} \circ \mathbb{P} + e^{-\mathfrak{i}s} \circ \mathbb{Q} = (P + p_z\mathfrak{i} + q_y\mathfrak{j} + q_x\mathfrak{k}) \circ e^{\mathfrak{i}s} + (Q + q_z\mathfrak{i} + p_y\mathfrak{j} + p_x\mathfrak{k}) \circ e^{-\mathfrak{i}s} \end{aligned} \tag{1.110}$$

where $\mathbb{P}$ and $\mathbb{Q}$ are *arbitrary quaternions* (the last equality only shows that post-multiplying by two general quaternions is equivalent to pre-multiplying by two different, but still fully general quaternions). Note that, for this purpose, we need to employ a (constant) gauge factor again.

Proof. We first define

$$\mathbb{S} \equiv e^{-\mathfrak{i}s_\mathrm{p}} \circ (A\cos\alpha - \mathfrak{i}A\,\gamma\sin\alpha + \mathfrak{k}B\sin\alpha) \tag{1.111}$$

and

$$\mathbb{T} \equiv e^{\mathfrak{i}s_\mathrm{p}} \circ (B\cos\alpha + \mathfrak{j}A\,\gamma\cos\alpha + \mathfrak{k}A\sin\alpha) \tag{1.112}$$

Note that since (1.110) can be written as

$$(e^{\mathfrak{i}s} \circ \mathbb{S} + e^{-\mathfrak{i}s} \circ \mathbb{T}) \circ \mathbb{R} \tag{1.113}$$

we get: $\mathbb{S} \circ \overline{\mathbb{T}} = \mathbb{P} \circ \overline{\mathbb{Q}}$, $\mathbb{S} \circ \overline{\mathbb{S}} = \mathbb{P} \circ \overline{\mathbb{P}}$ and $\mathbb{T} \circ \overline{\mathbb{T}} = \mathbb{Q} \circ \overline{\mathbb{Q}}$.

This implies that

$$\mathbb{P}\circ\overline{\mathbb{Q}} = AB\cos 2s_{\rm p} - \mathfrak{i}\, AB\sin 2s_{\rm p} - \mathfrak{j}\,\gamma A^2 - \frac{\mathfrak{k}}{2}(1-\gamma^2)A^2\sin 2\alpha + \frac{\mathfrak{k}}{2}B^2\sin 2\alpha \tag{1.114a}$$

$$\mathbb{P}\circ\overline{\mathbb{P}} + \mathbb{Q}\circ\overline{\mathbb{Q}} = A^2 + B^2 + A^2\gamma^2 \tag{1.114b}$$

and

$$\mathbb{P}\circ\overline{\mathbb{P}} - \mathbb{Q}\circ\overline{\mathbb{Q}} = [A^2(1-\gamma^2) - B^2]\cos 2\alpha \tag{1.114c}$$

The last three equations (the first being quaternionic, giving us four pieces of information) can be easily solved for $s_{\rm p}$, α, A, B and γ (in that order). $\mathbb{R}$ can then be computed from either $\overline{\mathbb{S}}\circ\mathbb{P}/(\mathbb{S}\circ\overline{\mathbb{S}})$ or $\overline{\mathbb{T}}\circ\mathbb{Q}/(\mathbb{T}\circ\overline{\mathbb{T}})$. The last two quantities are identical (which follows by post-multiplying them by $\overline{\mathbb{Q}}$) and they meet the $\mathbb{R}\circ\overline{\mathbb{R}} \equiv 1$ condition.

By explicitly finding their individual parameters, we have thus shown that $\mathbb{S}\circ\mathbb{R}$ and $\mathbb{T}\circ\mathbb{R}$ can have each an arbitrary quaternionic value. □

Obviously, in the trial solution (1.105), $\mathcal{S}_0 = \mathfrak{j}\,\gamma$, $a = A^2 + B^2$, and $\beta = A/B$, where A, B and γ are the parameters of the previous proof ($s_{\rm p}$ and $\mathbb{R}$ remain unchanged). We have chosen to eliminate the α-parameter (by making it equal to zero), as a part of our new gauge.

Also, from now on, it will be convenient to explicitly separate $\mathcal{S}_0 \equiv \mathfrak{j}\, b$ from the rest of $\mathcal{S}(z)$, writing the trial solution in the following form

$$\sqrt{\frac{a}{1+\beta^2}}\,q\circ\left(1+\frac{\beta}{z}+\mathcal{D}(z)+\mathfrak{k}\circ\frac{\mathfrak{i}\,b+\mathcal{S}(z)}{1+\beta\, z}\right)\circ\mathbb{R} \tag{1.115}$$

where $\mathcal{S}(z)$ is now a power series without the $\mathcal{S}_{-1}$, $\mathcal{S}_0$ and $\mathcal{S}_1$ terms.

How do we know the this form of a trial solution is just about right, to enable us to solve (1.29) in a unique manner? Well, at this point, we don't - this can be established only by constructing such a solution (in the next chapter).

1.4.2 Orbital elements

To construct the actual solution, we have to substitute (1.115) into (1.29), and solve for (1.106) and (1.107) by making the overall coefficient of each power of q (note that only *odd* powers will be encountered) equal to zero.

In a single step, this can be achieved only with terms proportional to ε, implying that the solution must be build in an iterative manner. Furthermore, we can be successful only by allowing a, β, $s_{\rm p}$ and the three parameters of $\mathbb{R}$ (the Euler angles) to be slowly varying functions of s; this is similar to solving $y'' + y = \varepsilon\sin x$ by $y = A(x)\sin x + B(x)\cos x$. We thus have to add a', β', $s_{\rm p}'$, ϕ', θ' and ψ' to the coefficients of $\mathcal{D}(z)$ and $\mathcal{S}(z)$, to complete the list of our unknowns.

In general, there are infinitely many odd powers of q; eliminating them all results in an infinite set of ordinary, linear equations - solving these is a task of the next chapter.

1.4.3 Conclusion

So, after all this (rather considerable) effort, we have delineated a procedure for converting a simple looking second-order differential equation (1.27) for three unknown functions $\mathbf{r}_x(t)$, $\mathbf{r}_y(t)$ and $\mathbf{r}_z(t)$ into six first-order differential equations for the orbital elements. The obvious question is: does this constitute a worthwhile simplification of the original problem?

The answer is a very resounding 'yes', for the following reasons. Firstly, our solution often leads to a simple, analytic answer, albeit expanded in terms of ε. Secondly, even when such a simplification is too difficult, one can always easily resort to constructing a valuable semi-analytic solution. Finally, when the perturbing force does not explicitly depend on t, numerical integration of the new set of differential equations can proceed hundreds of times faster than in case of (1.27), as they turn out to be AUTONOMOUS (the right hand sides are t independent as well).

CHAPTER 2

Iterative solution of perturbed problem

Abstract

Starting with the unperturbed solution, we now build a general, arbitrarily accurate analytic solution to the *perturbed* Kepler problem, assuming that the perturbation is proportional to a *small* parameter denoted ε.[1] This is done in an iterative manner, where the crucial step is constructing an ε-accurate solution; subsequent iterations then follow the same pattern. The chapter involves a lot of tedious but inevitable algebra, and needs to be read in detail only by those who want to understand the derivation of all key formulas. Anyone interested only in applications of the technique should proceed directly to the chapter's last section (the technique's summary).

In terms of actually solving (1.29), it is convenient to treat separately two distinct cases:

1. The AUTONOMOUS case, in which the perturbing force $\varepsilon\mathbf{f}$ has no *explicit* time dependence.
2. The case of $\varepsilon\mathbf{f}$ being an explicit (we will assume periodic) function of time.

In either case, we find it convenient to work in the orbit's KEPLER FRAME, introduced in (1.16); we have yet to elaborate on some of the details.

2.1 KEPLER FRAME

As we already know, this is a special coordinate system, having the same origin as the old (inertial) one, but whose x-direction always points towards the orbit's apocenter, and the whose z-axis coincides with the orbit's normal (it is thus the old coordinate system rotated by $\mathbb{R}$).

Similarly to (1.16), $\mathbf{r}_\text{o} = \mathbb{R} \circ \mathbf{r} \circ \overline{\mathbb{R}}$ facilitates the transformation of a vector $\mathbf{r}$ from the original, inertial set of coordinates, to its Kepler-frame representation $\mathbf{r}_\text{o}$. For the quaternion function $\mathbb{U}$ of (1.28a), this corresponds to $\mathbb{U}_\text{o} = \mathbb{U} \circ \overline{\mathbb{R}}$ ($\mathbb{U}$ thus transforms as a kind of half-vector, called SPINOR), implying that

$$\mathbb{U} = \mathbb{U}_\text{o} \circ \mathbb{R} \tag{2.1}$$

and therefore

$$\mathbb{U}' = \mathbb{U}_\text{o}' \circ \mathbb{R} + \mathbb{U}_\text{o} \circ \mathbb{R}' \equiv (\mathbb{U}_\text{o}' + \mathbb{U}_\text{o} \circ \frac{\mathbf{Z}_\text{o}}{2}) \circ \mathbb{R} \tag{2.2}$$

and

$$\mathbb{U}'' = (\mathbb{U}_\text{o}'' + \mathbb{U}_\text{o}' \circ \mathbf{Z}_\text{o} + \mathbb{U}_\text{o} \circ \frac{\mathbf{Z}_\text{o}'}{2} + \mathbb{U}_\text{o} \circ \frac{\mathbf{Z}_\text{o}^2}{4}) \circ \mathbb{R} \tag{2.3}$$

where $\mathbf{Z}_\text{o}$ (see Eq. 1.16) is the angular velocity of the Kepler's frame rotation, expressed in the same Kepler frame.

[1]This chapter is based on [5], [56], [40]–[42] and [46].

Jan Vrbik

Note that Γ (see Eq. 1.30), which is a scalar and therefore *invariant* under rotation (thus,we don't need the redundant Kepler-frame subscript), can now be computed by

$$\Gamma = \overline{\mathbb{U}}_o \circ \mathfrak{k} \circ \mathbb{U}'_o - \overline{\mathbb{U}}'_o \circ \mathfrak{k} \circ \mathbb{U}_o + \frac{\mathbf{r}_o \circ \mathbf{Z}_o + \mathbf{Z}_o \circ \mathbf{r}_o}{2} = 2\mathrm{Re}\left(\overline{\mathbb{U}}_o \circ \mathfrak{k} \circ \mathbb{U}'_o + \frac{\mathbf{r}_o \circ \mathbf{Z}_o}{2}\right) \tag{2.4}$$

Proof.

$$\begin{aligned}\Gamma &= 2\mathrm{Re}(\overline{\mathbb{U}} \circ \mathfrak{k} \circ \mathbb{U}') = 2\mathrm{Re}\left(\overline{\mathbb{R}} \circ \overline{\mathbb{U}}_o \circ \mathfrak{k} \circ (\mathbb{U}'_o + \mathbb{U}_o \circ \frac{\mathbf{Z}_o}{2}) \circ \mathbb{R}\right) \\ &= 2\mathrm{Re}\left(\overline{\mathbb{U}}_o \circ \mathfrak{k} \circ (\mathbb{U}'_o + \mathbb{U}_o \circ \frac{\mathbf{Z}_o}{2})\right) = 2\mathrm{Re}\left(\overline{\mathbb{U}}_o \circ \mathfrak{k} \circ \mathbb{U}'_o + \frac{\mathbf{r}_o \circ \mathbf{Z}_o}{2}\right)\end{aligned} \tag{2.5}$$

□

When the perturbing force lies within the orbit's plane (the so called *planar* assumption), $\mathbb{U}_o$ (and, consequently, $\overline{\mathbb{U}}_o$, $\mathbb{U}'_o$ and $\mathbf{r}_o$) will have only $\mathfrak{k}$ and $\mathfrak{j}$ components, whereas $\mathbf{Z}_o$ will be perpendicular to the orbit's plane, thus having only an $\mathfrak{i}$ component. This implies that $\overline{\mathbb{U}}_o \circ \mathfrak{k} \circ \mathbb{U}'_o + \mathbf{r}_o \circ \mathbf{Z}_o/2$ has only $\mathfrak{k}$ and $\mathfrak{j}$ components, and the corresponding Γ is automatically equal to zero.

2.1.1 The main equation

In the same manner, (1.29) can be converted to Kepler's frame by postmultiplying it by $\overline{\mathbb{R}}$, which yields

$$\begin{aligned}&2\left(\mathbb{U}''_o + \mathbb{U}'_o \circ \mathbf{Z}_o + \mathbb{U}_o \circ \frac{\mathbf{Z}'_o}{2} + \mathbb{U}_o \frac{\mathbf{Z}_o^2}{4}\right) - \left(2\left|\mathbb{U}'_o + \mathbb{U}_o \circ \frac{\mathbf{Z}_o}{2}\right|^2 - 4a\right)\frac{\mathbb{U}_o}{r} + 2\mathfrak{k} \circ \left(\mathbb{U}'_o + \mathbb{U}_o \circ \frac{\mathbf{Z}_o}{2}\right)\frac{\Gamma}{r} \\ &+\mathfrak{k} \circ \mathbb{U}_o \left(\frac{\Gamma}{r}\right)' - \left(\mathbb{U}'_o + \mathbb{U}_o \circ \frac{\mathbf{Z}_o}{2} + \mathfrak{k} \circ \mathbb{U}_o \frac{\Gamma}{2r}\right)\frac{a'}{a} + 4\frac{a}{\mu}\varepsilon \mathbb{U}_o \circ \mathbf{r}_o \circ \mathbf{f}_o = 0\end{aligned} \tag{2.6}$$

This is obvious from (2.1)-(2.3). Note that Γ is now computed based on (2.5).

2.2 AUTONOMOUS CASE

In this section we assume that $\varepsilon\mathbf{f}$ is a function of $\mathbf{r}$ (in some cases also of $\dot{\mathbf{r}}$, but *not* of time t). The solution to (2.6) will be built in an iterative manner: first we construct an ε-accurate solution which, when substituted back into (1.29), enables us to construct an ε^2-accurate solution, in essentially the same manner, etc.

2.2.1 Method of solution

We will now concentrate on the specifics of constructing an ε^n-accurate solution to (2.6), assuming that an ε^{n-1}-accurate solution has already been found. Starting with $n = 1$ and the ε^0-accurate (i.e. unperturbed) solution, this will enable us to extend it, ITERATIVELY, to an *arbitrarily accurate* analytic solution.

First, we symbolically expand the coefficients of $\mathcal{D}(z)$ and $\mathcal{S}(z)$, b, a', β', s'_p and $\mathbf{Z}_o$ (equivalent to ϕ', θ' and ψ') — these are the unknowns of trial solution (1.115) — in powers of ε, starting with linear terms, e.g.

$$\begin{aligned}\mathcal{D}_i &= \mathcal{D}_i^{(1)}\varepsilon + \mathcal{D}_i^{(2)}\varepsilon^2 + \cdots + \mathcal{D}_i^{(n)}\varepsilon^n \\ &\vdots \\ a' &= a'_{(1)}\varepsilon + a'_{(2)}\varepsilon^2 + \cdots + a'_{(n)}\varepsilon^n + \\ &\vdots\end{aligned} \tag{2.7}$$

etc., where the first $n-1$ coefficients of each expansion are already specific functions of a, β, ϕ, θ and ψ (they represent the ε^{n-1}-accurate solution, assumed known), but the last, n^{th} coefficient is yet to be found (these are the unknowns). Note that differentiation (with respect to s) increases 'smallness' of each of these coefficients by one power of ε. This happens automatically when differentiating the 'known' coefficients; when differentiating the 'unknown' terms such as $a'_{(n)}\varepsilon^n$, we get a ε^{n+1}-proportional quantity (which can be discarded, as we learn shortly).

We then substitute the corresponding trial solution (1.105) into the left had side of (2.6) and expand it in powers of ε, keeping only ε^n-proportional terms (discarding ε^{n+1}, $\varepsilon^{n+2}, \cdots$ - note that terms with ε-power *lower* than n must all cancel out). The result is premultiplied by a conjugate of the unperturbed solution (i.e. by $\mathcal{U}_0^*$) and expanded in powers (both positive and negative) of z. Making the coefficients of all powers of z equal to zero yields a set of *ordinary, linear* equations for the unknowns. Solving these is not quite so simple, as the number of unknowns is infinite, and the equations are quaternionic. Luckily, one can partially de-couple these, replacing each quaternionic equation by *two complex* ones. It is then possible to solve for all unknowns.

2.2.1.1 Equation de-coupling

To carry out the plan delineated above, we substitute the ε^{n-1}-accurate solution (*without* the unknown terms, which will be considered separately) into the left hand side of (2.6), evaluating it to the ε^n-accuracy (only ε^n-proportional terms remain) and calling the resulting (quaternionic) expression $\mathbb{F}_o^{(n)}$.

We then define two complex quantities

$$\mathcal{Q}_{(n)}(z) \equiv \frac{-\mathrm{Cx}(\mathcal{U}_0^* \circ \mathbb{F}_o^{(n)})}{2r_0(1+\beta z)\varepsilon^n} \tag{2.8}$$

and

$$\mathcal{W}_{(n)}(z) \equiv \frac{\mathrm{Cx}(\mathcal{U}_0^* \circ \mathfrak{k} \circ \mathbb{F}_o)}{r_0\varepsilon^n} \tag{2.9}$$

(these equal to $-2a\mathrm{Cx}(\mathbf{r}_0 \circ \mathbf{f}_0)/\mu/(1+\beta z)$ and $-4ar_0\mathrm{Cx}(\mathbf{f}_0)/\mu$, respectively, in the important case of the first iteration), where

$$\mathcal{U}_0 = \sqrt{\frac{a}{1+\beta^2}}\left(q+\frac{\beta}{q}\right) \tag{2.10}$$

$$r_0 \equiv \mathcal{U}_0^*\mathcal{U}_0 = \frac{a}{1+\beta^2}(1+\beta z)(1+\frac{\beta}{z}) \tag{2.11}$$

and $\mathrm{Cx}(\mathbb{A})$ extracts the COMPLEX part (real and $\mathfrak{i}$ components) of a quaternionic argument (discarding its $\mathfrak{j}$ and $\mathfrak{k}$ components). Be careful to differentiate between the 'o' subscript (denoting Kepler-frame quantity), and the '0' subscript, reserved for the Kepler-frame *unperturbed* solution.

Note that $\mathcal{W}_{(n)}(z)^* = -\mathcal{W}_{(n)}(z)$. This remains true at each step of the iteration procedure, and it is this important property of $\mathcal{W}_{(n)}(z)$ that enabled us in the last chapter, somehow clairvoyantly, to choose the new gauge (1.108).

Proof. We can show that the numerator of (2.9) - the corresponding denominator is always real - can be written in the following manner:

$$\mathrm{Cx}(\mathcal{U}_0^* \circ \mathfrak{k} \circ \mathbb{F}_o^{(n)}) \doteqdot \mathrm{Cx}(\overline{\mathbb{U}}_o \circ \mathfrak{k} \circ \mathbb{F}_o^{(n)}) = \mathrm{Cx}(\mathbb{R} \circ \overline{\mathbb{U}} \circ \mathfrak{k} \circ \mathbb{F}^{(n)} \circ \overline{\mathbb{R}}) \tag{2.12}$$

where $\mathbb{F}^{(n)}$ is the left hand side of (1.29) and $\doteqdot$ implies that only ε^n-accurate terms are considered.

The first equality rests on the fact that $\mathbb{F}_o^{(n)}$ is proportional to ε^n, thus the difference between $\mathcal{U}_0^*$ and $\overline{\mathbb{U}}_o$ contributes only to higher-order terms; the second equality follows from the definitions of $\overline{\mathbb{U}}_o$ and $\mathbb{F}_o^{(n)}$.

All we need to show is that $\overline{\mathbb{U}} \circ \mathfrak{k} \circ \mathbb{F}^{(n)}$ is a vector (implying that $\mathbb{R} \circ \overline{\mathbb{U}} \circ \mathfrak{k} \circ \mathbb{F}^{(n)} \circ \overline{\mathbb{R}}$ is a vector, further implying that $\mathcal{W}_{(n)}(z)$ has no real part). This (being a vector) is obvious for the $(2|\mathbb{U}'|^2 - 4a)\mathbb{U}/r$ and $4a\varepsilon\mathbb{U}/\mu \circ \mathbf{r} \circ \mathbf{f}$ terms of (1.29); it follows easily for

$$\overline{\mathbb{U}} \circ \mathfrak{k} \circ \left(\mathbb{U}' + \mathfrak{k} \circ \mathbb{U}\frac{\Gamma}{2r}\right)\frac{a'}{a} = (\overline{\mathbb{U}} \circ \mathfrak{k} \circ \mathbb{U}' + \overline{\mathbb{U}}' \circ \mathfrak{k} \circ \mathbb{U})\frac{a'}{2a} \tag{2.13}$$

(its conjugate just changes sign), and for the remaining

$$2\overline{\mathbb{U}} \circ \mathfrak{k} \circ \mathbb{U}'' - 2\overline{\mathbb{U}} \circ \mathbb{U}'\frac{\Gamma}{r} - r\left(\frac{\Gamma}{r}\right)' = \overline{\mathbb{U}} \circ \mathfrak{k} \circ \mathbb{U}'' + \overline{\mathbb{U}}'' \circ \mathfrak{k} \circ \mathbb{U} + (\overline{\mathbb{U}}' \circ \mathbb{U} - \overline{\mathbb{U}} \circ \mathbb{U}')\frac{\Gamma}{r} \tag{2.14}$$

(again, take the conjugate), since

$$\left(\frac{\Gamma}{r}\right)' = \frac{\overline{\mathbb{U}} \circ \mathfrak{k} \circ \mathbb{U}'' - \overline{\mathbb{U}}'' \circ \mathfrak{k} \circ \mathbb{U}}{r} - \frac{\Gamma}{r^2}(\overline{\mathbb{U}}' \circ \mathbb{U} + \overline{\mathbb{U}} \circ \mathbb{U}' \tag{2.15}$$

□

Once $\mathcal{Q}_{(n)}(z)$ and $\mathcal{W}_{(n)}(z)$ have been computed, we expand them in a Laurent series of all (positive and negative) powers in z, thus

$$\mathcal{Q}_{(n)}(z) \equiv \cdots + \frac{\mathcal{Q}_{-2}}{z^2} + \frac{\mathcal{Q}_{-1}}{z} + \mathcal{Q}_0 + \mathcal{Q}_1 z + \mathcal{Q}_2 z^2 + \cdots \tag{2.16}$$

and

$$\mathcal{W}_{(n)}(z) = \cdots + \frac{-\mathcal{W}_2^*}{z^2} + \frac{-\mathcal{W}_1^*}{z} + \mathcal{W}_0 + \mathcal{W}_1 z + \mathcal{W}_2 z^2 + \cdots . \tag{2.17}$$

where $\mathcal{W}_0$ is purely imaginary. In some cases (of a simple perturbing force $\varepsilon\mathbf{f}$), the series expansion may be achieved purely algebraically; in others, it may require contour integration (with respect to z - the orbital elements are considered fixed). Note that this expansion must be valid on the unit circle; therefore, the contour integration has to account for all singularities inside this circle (normally, that would include $z = 0$ and $z = -\beta$).

2.2.1.2 'Unknown' terms

When substituting the *complete* trial solution into the left hand side of (2.6), we get $\mathbb{F}_\text{o}^{(n)}$ *plus* terms (so far omitted in the computation of $\mathbb{F}_\text{o}^{(n)}$) proportional to the *unknowns*. To obtain these, we differentiate the trial solution, keeping *only* the unperturbed and ε^n-proportional terms of (2.7) - the rest of them have already been accounted for, and incorporated into $\mathbb{F}_\text{o}^{(n)}$ (note that extra derivatives of the unknowns, such as $a''_{(n)}$, are of the ε^{n+1} type and therefore discarded). This yields

$$\mathbb{U}_\text{o} = \sqrt{\frac{a}{1+\beta^2}}\left(q + \frac{\beta}{q} + \cdots + q\varepsilon^n\mathcal{D}_{(n)}(z) + \mathfrak{k} \circ \frac{q\varepsilon^n\mathcal{S}_{(n)}(z)}{1+\beta z} + \mathfrak{j} \circ \frac{q\varepsilon^n b_{(n)}}{1+\beta z}\right) \tag{2.18}$$

which needs to be differentiated once

$$\begin{aligned}
&\mathbb{U}_\text{o}' = \\
&\sqrt{\frac{a}{1+\beta^2}}\left(q(\mathrm{i} - \mathrm{i}\varepsilon^n s'_{\mathrm{p}(n)}) - \frac{\beta}{q}(\mathrm{i} - \mathrm{i}\varepsilon^n s'_{\mathrm{p}(n)})\right) + \sqrt{\frac{a}{1+\beta^2}}\left(q + \frac{\beta}{q}\right)\frac{\varepsilon^n a'_{(n)}}{2a} + \sqrt{\frac{a}{1+\beta^2}}\left(\frac{1}{q} - \beta q\right)\frac{\varepsilon^n \beta'_{(n)}}{1+\beta^2} \\
&+\varepsilon^n\sqrt{\frac{a}{1+\beta^2}}\left(\mathrm{i}q\mathcal{D}_{(n)}(z) + 2\mathrm{i}qz\frac{\mathrm{d}\mathcal{D}_{(n)}(z)}{\mathrm{d}z} - \mathfrak{k} \circ q\frac{1-\beta z}{(1+\beta z)^2}b_{(n)} + \mathfrak{j} \circ q\frac{1-\beta z}{(1+\beta z)^2}\mathcal{S}_{(n)}(z)\right. \\
&\left. + 2\mathfrak{j} \circ q\frac{z}{1+\beta z}\frac{\mathrm{d}\mathcal{S}_{(n)}(z)}{\mathrm{d}z}\right)
\end{aligned} \tag{2.19}$$

and again

$$\mathbb{U}_\mathrm{o}'' =$$
$$-\sqrt{\frac{a}{1+\beta^2}}\left(q(1-2\varepsilon^n s'_{\mathrm{p}(n)}) + \frac{\beta}{q}(1-2\varepsilon^n s'_{\mathrm{p}(n)})\right) + \sqrt{\frac{a}{1+\beta^2}}\mathbf{i}\left(q-\frac{\beta}{q}\right)\frac{\varepsilon^n a'_{(n)}}{a}$$
$$-\sqrt{\frac{a}{1+\beta^2}}2\mathbf{i}\left(\frac{1}{q}+\beta q\right)\frac{\varepsilon^n \beta'_{(n)}}{1+\beta^2} + \sqrt{\frac{a}{1+\beta^2}}\varepsilon^n\left(-q\mathcal{D}_{(n)}(z) - 8qz\frac{\mathrm{d}\mathcal{D}_{(n)}(z)}{\mathrm{d}z} - 4qz^2\frac{\mathrm{d}^2\mathcal{D}_{(n)}(z)}{\mathrm{d}z^2} - \mathbf{j}\circ q\times\right.$$
$$\left.\frac{1-6\beta z+\beta^2 z^2}{(1+\beta z)^3}b_{(n)} - \mathfrak{k}\circ q\frac{1-6\beta z+\beta^2 z^2}{(1+\beta z)^3}\mathcal{S}_{(n)}(z) - 8\mathfrak{k}\circ q\frac{z}{(1+\beta z)^2}\frac{\mathrm{d}\mathcal{S}_{(n)}(z)}{\mathrm{d}z} - 4\mathfrak{k}\circ q\frac{z^2}{1+\beta z}\frac{\mathrm{d}^2\mathcal{S}_{(n)}(z)}{\mathrm{d}z^2}\right) \tag{2.20}$$

where

$$\mathcal{D}_{(n)}(z) = \sum_{i\neq 0,1} D_i^{(n)} z^i \tag{2.21a}$$
$$\mathcal{S}_{(n)}(z) = \sum_{i\neq 0,1,-1} S_i^{(n)} z^i \tag{2.21b}$$

and z-differentiation considers the orbital elements constant.

Substituting this solution into (2.6) results in terms which either cancel out (the unperturbed part), have been already included in $\mathbb{F}_\mathrm{o}^{(n)}$, contribute to higher-than-n powers of ε, or are proportional to ε^n and one of our unknowns (only the last ones need to be considered). The left hand side of the equation thus reduces to

$$2\left(\mathbb{U}_\mathrm{o}'' + \mathcal{U}_0'\circ\varepsilon^n\mathbf{Z}_\mathrm{o}^{(n)}\right) - \left(2\mathbb{U}_\mathrm{o}'\circ\overline{\mathbb{U}}_\mathrm{o}' + \mathcal{U}_0\circ\frac{\varepsilon^n\mathbf{Z}_\mathrm{o}^{(n)}}{2}\circ\mathcal{U}_0'^* - \mathcal{U}_0'\circ\frac{\varepsilon^n\mathbf{Z}_\mathrm{o}^{(n)}}{2}\circ\mathcal{U}_0^* - 4a\right)\frac{\mathbb{U}_\mathrm{o}}{r} + 2\mathfrak{k}\circ\mathcal{U}_0'\frac{\varepsilon^n\Gamma_{(n)}}{r_0}$$
$$+\mathfrak{k}\circ\mathcal{U}_0\left(\frac{\varepsilon^n\Gamma_{(n)}}{r_0}\right)' - \mathcal{U}_0'\frac{\varepsilon^n a'_{(n)}}{a} \tag{2.22}$$

where Γ has been replaced by its 'unknown' part (note that Γ has no ε^0 term)

$$\varepsilon^n\Gamma_{(n)} = \frac{a\varepsilon^n}{1+\beta^2}\left(\frac{4(1-\beta^2)\left[b_{(n)} - \mathbf{i}\mathcal{S}_{(n)}(z)\right]}{(1+\beta z)(1+\frac{\beta}{z})} - \left[4\beta + (1+\beta^2)(z+\frac{1}{z})\right]\frac{Z_1^{(n)}}{2} + \mathbf{i}(z-\frac{1}{z})(1-\beta^2)\frac{Z_2^{(n)}}{2}\right) \tag{2.23}$$

In (2.22), terms containing the s-derivative or second power of $\varepsilon^n\mathbf{Z}_\mathrm{o}^{(n)}$ have been discarded (their contribution goes beyond ε^n). This is also the reason for disappearance of several other terms (including the perturbing one).

2.2.1.3 Complex part

Premultiplying (2.22) by $\overline{\mathbb{U}}_\mathrm{o}$ (equivalent to premultiplying it by $\mathcal{U}_0^*$) and discarding terms with no 'complex' (i.e. real or $\mathbf{i}$) part yields

$$2\overline{\mathbb{U}}_\mathrm{o}\circ\mathbb{U}_\mathrm{o}'' + 2\varepsilon^n Z_3^{(n)} - 2|\mathbb{U}_\mathrm{o}'|^2 + \mathcal{U}_0^*\mathcal{U}_0'\varepsilon^n Z_3^{(n)} - \mathcal{U}_0\mathcal{U}_0'^*\varepsilon^n Z_3^{(n)} + 4a - \mathcal{U}_0^*\mathcal{U}_0'\frac{\varepsilon^n a'_{(n)}}{a} \tag{2.24}$$

where

$$\mathcal{U}_0^* \mathcal{U}_0' = \frac{\mathrm{i}a}{1+\beta^2}(1+\beta z)(1-\frac{\beta}{z}) \tag{2.25}$$

and $\mathcal{U}_0 \mathcal{U}_0'^*$ is its complex conjugate.

Evaluating the complex part of the (2.24), and multiplying the answer by $(1+\beta^2)/(2a\varepsilon^n)$, yields

$$\mathrm{i}\left(1-\beta^2-\beta(z-\frac{1}{z})\right)\frac{a'_{(n)}}{2a} - \mathrm{i}(z+\frac{1}{z})\beta'_{(n)} - 4\mathrm{i}\frac{\beta}{1+\beta^2}\beta'_{(n)} - \left(2(1-\beta^2)+\beta(z-\frac{1}{z})\right)Z_3^{(n)}$$
$$+4(1+\beta^2)s'_{\mathrm{p}(n)} - 2\mathcal{D}_{(n)}(z) - 2\mathcal{D}^*_{(n)}(z) - 2z(5+3z\beta)\frac{\mathrm{d}\mathcal{D}_{(n)}(z)}{\mathrm{d}z} + 2(z-\beta)\frac{\mathrm{d}\mathcal{D}^*_{(n)}(z)}{\mathrm{d}z} - 4z(1+\beta z)\frac{\mathrm{d}^2\mathcal{D}_{(n)}(z)}{\mathrm{d}z^2} \tag{2.26}$$

Note that

$$\left(\frac{\mathrm{d}\mathcal{D}_{(n)}(z)}{\mathrm{d}z}\right)^* = -z^2\frac{\mathrm{d}\mathcal{D}^*_{(n)}(z)}{\mathrm{d}z} \tag{2.27}$$

The sum of (2.26) and of

$$\frac{1+\beta^2}{a}r_0(1+\beta z)\mathcal{Q}_{(n)}(z) = (1+\frac{\beta}{z})(1+\beta z)^2\mathcal{Q}_{(n)}(z) \tag{2.28}$$

represents the complete *complex* part of the left hand side of (2.6), premultiplied by $(1+\beta^2)\mathcal{U}_0^*/(2a\varepsilon^n)$. We can thus get a solution to $a'_{(n)}$, $\beta'_{(n)}$, $Z_3^{(n)}$, $s'_{\mathrm{p}(n)}$ and $\mathcal{D}_{(n)}(z)$ by making coefficients of all powers of z in this sum equal to zero. This requires a bit of algebra, but one can show that the answer is

$$a'_{(n)} = 2a\mathrm{Im}(\mathcal{Q}_0 - \beta\mathcal{Q}_{-1}) \tag{2.29a}$$

$$\beta'_{(n)} = -\frac{1+\beta^2}{4}\mathrm{Im}(\mathcal{Q}_1 + 3\beta\mathcal{Q}_0 + 3\mathcal{Q}_{-1} + \beta\mathcal{Q}_{-2}) \tag{2.29b}$$

$$Z_3^{(n)} = \frac{1}{4\beta}\times \tag{2.29c}$$
$$\mathrm{Re}\left(-(1+\beta^2)\mathcal{Q}_1 + \beta(1-3\beta^2)\mathcal{Q}_0 + (3-\beta^2)\mathcal{Q}_{-1} + \beta(1+\beta^2)\mathcal{Q}_{-2}\right)$$

$$s'_{\mathrm{p}(n)} - \frac{Z_3^{(n)}}{2} = \frac{1}{4(1+\beta^2)}\times\mathrm{Re}\left(\beta(2+\beta^2)\mathcal{Q}_1 + (1+\beta^2+3\beta^4)\mathcal{Q}_0 - \beta(1-2\beta^2)\mathcal{Q}_{-1} - \beta^4\mathcal{Q}_{-2}\right) \tag{2.29d}$$

and

$$\mathcal{D}_{(n)}(z) = -\frac{1}{4}\sum_{\substack{k=-\infty\\k\neq-1,0}}^{\infty}\left(\frac{\beta(k+\frac{1}{2})\mathcal{Q}_{k-1} + (k-\frac{1}{2})\mathcal{Q}_k + \frac{1}{2}\mathcal{Q}^*_{-k}}{k^2(k+1)}\right.$$
$$\left.+\frac{\beta^2(k+\frac{3}{2})\mathcal{Q}_k + \beta(k+\frac{1}{2})\mathcal{Q}_{k+1} - \frac{1}{2}\beta^2\mathcal{Q}^*_{-k-2}}{k(k+1)^2} - \frac{\frac{1}{2}\beta\mathcal{Q}^*_{-k-1}}{k^2(k+1)^2}\right)z^k \tag{2.29e}$$

Note that $\mathcal{Q}^*_{-k}$ denotes the complex conjugate of the coefficient of z^{-k} in the expansion of $\mathcal{Q}_{(n)}(z)$, which is the same as the coefficient of z^k of in the expansion of $\mathcal{Q}^*_{(n)}(z)$.

Proof. First, one can show (by collecting Q_k and Q_k^*-proportional terms, individually, for each integer k) that

$$\begin{aligned}&(1+\frac{\beta}{z})(1+\beta z)^2\mathcal{Q}_{(n)}(z)-2\mathcal{D}_{(n)}(z)-2\mathcal{D}^*_{(n)}(z)-2z(5+3z\beta)\frac{\mathrm{d}\mathcal{D}_{(n)}(z)}{\mathrm{d}z}+2(z-\beta)\frac{\mathrm{d}\mathcal{D}^*_{(n)}(z)}{\mathrm{d}z}\\ &-4z(1+\beta z)\frac{\mathrm{d}^2\mathcal{D}_{(n)}(z)}{\mathrm{d}z^2} = \beta\left(\beta+\frac{1+\beta^2}{4z}\right)Q_{-2}-\beta\frac{1+\beta^2}{4}zQ^*_{-2}+\left(\beta(2+\beta^2)+\frac{3-\beta^2}{4z}+\beta^2 z\right)Q_{-1}\\ &-\frac{3}{4}(1+\beta^2)zQ^*_{-1}+\left(1+2\beta^2+\frac{\beta}{z}-\beta\frac{1-3\beta^2}{4}z\right)Q_0-3\beta\frac{1+\beta^2}{4z}Q^*_0+\left(\beta+\frac{1+\beta^2}{4}z\right)Q_1-\frac{1+\beta^2}{4z}Q^*_1\end{aligned} \tag{2.30}$$

which has removed all powers of z, with the exception of z^{-1}, z^0 and z^1. Adding

$$\mathrm{i}\left(1-\beta^2-\beta(z-\frac{1}{z})\right)\frac{a'_{(n)}}{2a}-\mathrm{i}(z+\frac{1}{z})\beta'_{(n)}-4\mathrm{i}\frac{\beta}{1+\beta^2}\beta'_{(n)}-\left(2(1-\beta^2)+\beta(z-\frac{1}{z})\right)Z_3^{(n)}+4(1+\beta^2)s'_{\mathrm{p}(n)} \tag{2.31}$$

to the last expression must eliminate the remaining three coefficients. This yields three complex (six real) *linear* equations for four real unknowns, $a'_{(n)}$, $\beta'_{(n)}$, $Z_3^{(n)}$ and $s'_{\mathrm{p}(n)}$ (they de-couple into two sets of three real equations for $a'_{(n)}$ and $\beta'_{(n)}$, and for $Z_3^{(n)}$ and $s'_{\mathrm{p}(n)}$, respectively). Luckily, even though overdetermined, each of these sets has a unique solution, resulting in (2.29), as one can easily verify. □

Note that, in the cases of more complicated perturbations, the easiest way to find a given combination of $\mathcal{Q}_i$ coefficients is by contour integration (which assumes the orbital parameters to be fixed), for example

$$a'_{(n)} = 2a\ \mathrm{Im}\left[\oint_{C_0}(1-\beta z)\mathcal{Q}_{(n)}(z)\frac{\mathrm{d}z}{2\pi\mathrm{i}z}\right] \tag{2.32}$$

and

$$\begin{aligned}8\mathcal{D}_{(n)}(z) = &\oint_{C_0}\Big[-[3(1+\beta^2)+\beta(u+\frac{1}{u})][\mathsf{G}_1(\frac{z}{u})+\frac{u}{z}\mathsf{G}_1(\frac{u}{z})]+(1-\beta u)[\mathsf{G}_2(\frac{z}{u})+\frac{u}{z}\mathsf{G}_3(\frac{u}{z})]\\ &\quad+\beta(\beta-\frac{1}{u})[\mathsf{G}_3(\frac{z}{u})+\frac{u}{z}\mathsf{G}_2(\frac{u}{z})]\Big]\ \mathcal{Q}_{(n)}(u)\frac{\mathrm{d}u}{2\pi\mathrm{i}u}\\ &+\oint_{C_0}\Big[(1-\frac{\beta}{u})^2[\mathsf{G}_1(\frac{z}{u})+\frac{u}{z}\mathsf{G}_1(\frac{u}{z})]-(1-\frac{\beta}{u})[\mathsf{G}_2(\frac{z}{u})+\frac{u}{z}\mathsf{G}_3(\frac{u}{z})]\\ &\quad+\frac{\beta}{u}(1-\frac{\beta}{u})[\mathsf{G}_3(\frac{z}{u})+\frac{u}{z}\mathsf{G}_2(\frac{u}{z})]\Big]\ \mathcal{Q}^*_{(n)}(u)\frac{\mathrm{d}u}{2\pi\mathrm{i}u}\end{aligned} \tag{2.33}$$

where

$$\mathsf{G}_1(z) \equiv \sum_{n=1}^{\infty}\frac{z^n}{n(n+1)} = 1+\frac{1-z}{z}\ln(1-z) \tag{2.34a}$$

$$\mathsf{G}_2(z) \equiv \sum_{n=1}^{\infty}\frac{z^n}{n^2} = \mathrm{di}\log(1-z) \tag{2.34b}$$

$$\mathsf{G}_3(z) \equiv \sum_{n=1}^{\infty} \frac{z^n}{(n+1)^2} = \frac{\operatorname{di}\log(1-z)}{z} - 1 \tag{2.34c}$$

(the last three functions are analytic inside the unit circle, and continuous including the boundary).

2.2.1.4 $\mathfrak{j}$-$\mathfrak{k}$ *part*

Similarly, premultiplying (2.22) by $\mathbf{r}_0 \circ \overline{\mathbb{U}}_\circ / r_0$ - equivalent to premultiplying it by $\mathcal{U}_0 \circ \mathfrak{k}$, but more convenient, since it readily eliminates the second term of (2.6) - and discarding terms which contribute beyond the ε^n accuracy or are not 'complex', we get

$$\frac{2\mathbf{r}_0}{r_0} \circ \overline{\mathbb{U}}_\circ \circ \left(\mathbb{U}_\circ'' + \mathcal{U}_0' \circ \varepsilon^n \mathbf{Z}_\circ^{(n)} \right) - 2\mathcal{U}_0^* \mathcal{U}_0' \frac{\varepsilon^n \Gamma_{(n)}}{r_0} - r_0 \left(\frac{\varepsilon^n \Gamma_{(n)}}{r_0} \right)' \tag{2.35}$$

This can be simplified (discarding more such terms) to

$$\frac{2\mathbf{r}_0}{r_0} \circ \overline{\mathbb{U}}_\circ \circ \mathbb{U}_\circ'' - 2\mathcal{U}_0^* \mathcal{U}_0'^* \varepsilon^n (Z_1 + \mathfrak{i} Z_2) + (\mathcal{U}_0 \mathcal{U}_0'^* - \mathcal{U}_0^* \mathcal{U}_0') \frac{\varepsilon^n \Gamma_{(n)}}{r_0} - \varepsilon^n \Gamma_{(n)}' \tag{2.36}$$

which, when multiplied by $(1+\beta^2)/a/\varepsilon^n$ and stripped of the last remaining $\mathfrak{j}$-$\mathfrak{k}$ components, yields

$$\mathfrak{i} \frac{(1-\beta^2)\left(2\beta + z + \frac{1}{z}\right)\left(2 + \beta(z + \frac{1}{z})\right)}{(1+\beta z)(1+\frac{\beta}{z})} Z_1^{(n)} + \frac{(z - \frac{1}{z})\left(2 + 2\beta^4 + \beta(1+\beta^2)(z + \frac{1}{z})\right)}{(1+\beta z)(1+\frac{\beta}{z})} Z_2^{(n)}$$
$$-8\mathfrak{i} \frac{1+\beta^2}{(1+\beta z)(1+\frac{\beta}{z})} b_{(n)} - 8 \frac{1+\beta^2}{(1+\beta z)(1+\frac{\beta}{z})} \mathcal{S}_{(n)}(z) + 8 \frac{z + 2\beta + z\beta^2}{(1+\beta z)(1+\frac{\beta}{z})} \frac{\mathrm{d}\mathcal{S}_{(n)}(z)}{\mathrm{d}z} + 8z^2 \frac{\mathrm{d}^2 \mathcal{S}_{(n)}(z)}{\mathrm{d}z^2} \tag{2.37}$$

Note that

$$\left(\frac{\mathrm{d}\mathcal{S}_{(n)}(z)}{\mathrm{d}z} \right)^* = z^2 \frac{\mathrm{d}\mathcal{S}_{(n)}(z)}{\mathrm{d}z} \tag{2.38a}$$

$$\left(\frac{\mathrm{d}^2\mathcal{S}_{(n)}(z)}{\mathrm{d}z^2} \right)^* = -z^4 \frac{\mathrm{d}^2\mathcal{S}_{(n)}(z)}{\mathrm{d}z^2} - 2z^3 \frac{\mathrm{d}\mathcal{S}_{(n)}(z)}{\mathrm{d}z} \tag{2.38b}$$

The sum of (2.37) and

$$\frac{1+\beta^2}{a} r_0 \mathcal{W}_{(n)}(z) = (1 + \frac{\beta}{z})(1+\beta z)\mathcal{W}_{(n)}(z) \tag{2.39}$$

represents the complete complex part of the left hand side of (2.6) — premultiplied by $(1+\beta^2)\mathcal{U}_0^* \circ \mathfrak{k}/(2a\varepsilon^n)$ — and as such it must vanish. We can thus get a solution to $Z_1^{(n)}$, $Z_2^{(n)}$, $b_{(n)}$ and $\mathcal{S}_{(n)}(z)$, by making coefficients of all powers of z in this sum equal to zero. The answer is

$$Z_1^{(n)} = -\frac{1}{1-\beta^2} \operatorname{Im}\left(\frac{1+\beta^2}{2} \mathcal{W}_1 + \beta \mathcal{W}_0 \right) \tag{2.40a}$$

$$Z_2^{(n)} = -\frac{1}{2} \operatorname{Re}(\mathcal{W}_1) \tag{2.40b}$$

$$b_{(n)} = \frac{1}{8} \operatorname{Im}\left[\frac{(1-\beta^2)^2 \mathcal{W}_0 + 2\beta^2 \mathcal{W}_2}{1+\beta^2} + \beta \mathcal{W}_1 \right] \tag{2.40c}$$

$$\mathcal{S}_{(n)}(z) = -\frac{\mathfrak{i}}{4} \operatorname{Im}\left[\sum_{k=2}^{\infty} \left(\frac{\beta \mathcal{W}_{k-1}}{(k-1)k} + \frac{(1+\beta^2)\mathcal{W}_k}{k^2-1} + \frac{\beta \mathcal{W}_{k+1}}{k(k+1)} \right) z^k \right] \tag{2.40d}$$

Proof. Using (2.40d), we first evaluate (again, it is easier to organize this computation by collecting, individually, all $\mathcal{W}_k$-proportional terms, where k is a non-negative integer):

$$(1+\beta z)(1+\frac{\beta}{z})\mathcal{W}_{(n)}(z) - 8\frac{1+\beta^2}{(1+\beta z)(1+\frac{\beta}{z})}\mathcal{S}_{(n)}(z) + 8\frac{z+2\beta+z\beta^2}{(1+\beta z)(1+\frac{\beta}{z})}\frac{\mathrm{d}\mathcal{S}_{(n)}(z)}{\mathrm{d}z} + 8z^2\frac{\mathrm{d}^2\mathcal{S}_{(n)}(z)}{\mathrm{d}z^2}$$
$$= (1+\beta z)(1+\frac{\beta}{z})\mathcal{W}_0 + \frac{\frac{\beta^2}{z} + 2\beta(1+\beta^2) + (1+\beta^2+\beta^4)z + \frac{\beta}{2}(1+\beta^2)z^2}{(1+\beta z)(1+\frac{\beta}{z})}\mathcal{W}_1$$
$$-\frac{\beta^2 z + 2\beta(1+\beta^2) + \frac{1+\beta^2+\beta^4}{z} + \frac{\beta}{2z^2}(1+\beta^2)}{(1+\beta z)(1+\frac{\beta}{z})}\mathcal{W}_1^* + \frac{\beta^2}{(1+\beta z)(1+\frac{\beta}{z})}\mathcal{W}_2 - \frac{\beta^2}{(1+\beta z)(1+\frac{\beta}{z})}\mathcal{W}_2^* \quad (2.41)$$

These, added to the remaining part of (2.37) and multiplied by $(1+\beta z)(1+\beta/z)$, yield five linear, real equations (one for the purely-imaginary absolute term, and two each for the complex coefficients of z and z^2) for the remaining three unknowns $Z_1^{(n)}$, $Z_2^{(n)}$ and $b_{(n)}$.These again de-couple into three simple (linearly dependent) equations for $Z_1^{(n)}$ and $b^{(n)}$, and two (duplicate) equations for $Z_2^{(n)}$, which can be easily and uniquely solved, resulting in (2.40a)-(2.40c). □

Again, all of these can be conveniently evaluated by contour integration, including

$$\mathcal{S}_{(n)}(z) = -\frac{\mathrm{i}}{4}\mathrm{Im}\left\{\oint_{C_0}\left[\beta(z+\frac{1}{u})\mathsf{G}_1(\frac{z}{u}) - \beta\frac{z}{2u^2} + (1+\beta^2)\mathsf{G}_4(\frac{z}{u})\right]\mathcal{W}_{(n)}(u)\frac{\mathrm{d}u}{2\pi\mathrm{i}u}\right\} \quad (2.42)$$

where

$$\mathsf{G}_4(z) \equiv \sum_{n=2}^{\infty}\frac{z^n}{n^2-1} = \frac{z+2}{4} + \frac{1-z^2}{2z}\ln(1-z) \quad (2.43)$$

(also analytic).

2.2.1.5 Solution

To construct an ε^n-accurate solution, we still need to carry out the following steps:

1. Compute ϕ', θ' and ψ' based on the final expression for Z_1, Z_2 and Z_3, and (1.19).

2. Divide a', β', ϕ', θ' and ψ' by $2(1-s_\mathrm{p}')$, to convert them into d../dω derivatives (thus eliminating the non-physical s, including s_p); expand each answer to the ε^n-accuracy.

3. Solve the corresponding set of autonomous differential equations (for a, β, ϕ, θ and ψ as functions of ω) using an appropriate iteration technique (many of them are readily available), to the same ε^n-accuracy.

4. The resulting solution for $\mathbb{U}$ can be easily converted to the correspondingly accurate expressions for $\mathbf{r}$ and r.

5. Similarly divide (1.28b) by $2(1-s_\mathrm{p}')$ and integrate the right hand side in terms of ω, to express t in terms of ω (doing the reverse requires expanding the answer not only in powers of ε, but also in powers of β - see Keplers' equation; this enables us to express $\mathbf{r}$ and r as functions of t, instead of ω, if desired).

2.3 TIME-DEPENDENT PERTURBATIONS

When $\mathbf{f}$ is a function of not only $\mathbf{r}$ but also, explicitly, of time t, the situation becomes more complicated. To simplify our task, we will assume that the t-dependence is *periodic*, with an angular frequency Ω, which means that it can be expressed as a sum of term of the type $\mathbf{g}e^{i\Omega t}$, where $\mathbf{g}$ has no explicit t dependence. This covers most of the usual situations, and can be extended to other (such as t-proportional) perturbations by a limiting process.

2.3.1 Method of solution

The solution can still be obtained in an iterative manner. First, using the current, ε^{n-1}-accurate solution, one integrates (1.28b) to express t in terms of ω, z and the orbital parameters, thus

$$t = \sqrt{\frac{a_o^3}{\mu}}\left(\omega + \frac{\beta}{1+\beta^2}\frac{z-\frac{1}{z}}{i} + \chi\right) \tag{2.44}$$

where a_o is the average value of a, and χ is an auxiliary variable (its initial value is $t_0\sqrt{\mu/a_o^3}$). Note that

$$\chi' = \frac{2\sqrt{a}}{a_o^{3/2}}r - 2\left(1+\frac{\beta}{1+\beta^2}(z+\frac{1}{z})\right)(1-s_p') - \frac{1-\beta^2}{(1+\beta^2)^2}\frac{z-\frac{1}{z}}{i}\beta' \tag{2.45}$$

which is an $O(\varepsilon)$ quantity, since

$$\frac{2\sqrt{a}}{a_o^{3/2}}r - 2\left(1+\frac{\beta}{1+\beta^2}(z+\frac{1}{z})\right) = \frac{2a^{3/2}}{a_o^{3/2}}\left(\frac{r}{a} - 1 - \frac{\beta}{1+\beta^2}(z+\frac{1}{z})\right) + 2\left(\frac{a^{3/2}}{a_o^{3/2}} - 1\right)\left(1+\frac{\beta}{1+\beta^2}(z+\frac{1}{z})\right) \tag{2.46}$$

and can be set equal to zero to start the first iteration. Here, we have assumed that

$$\frac{a^{3/2}}{a_o^{3/2}} - 1 \tag{2.47}$$

is of the order of ε, even though it is more commonly of the $O(\varepsilon^3)$ type. But there are situations when (2.47), similarly to β, requires to be treated as 'small' independently of the ε expansion; in other cases it may behave as $O(\varepsilon^{1/2})$, $O(\varepsilon^{1/3})$, etc.

When (2.45) is treated as an $O(\varepsilon)$ quantity, differentiating it does *not* introduce the usual extra power of ε, implying that *two* rounds of the procedure are required to complete the *second* iteration (another two rounds are needed to reach ε^3 accuracy, and so on).

Using (2.44), $e^{i\Omega t}$ now becomes

$$e^{i\Lambda\omega}\cdot e^{\Lambda\frac{\beta}{1+\beta^2}(z-\frac{1}{z})}\cdot e^{i\Lambda\chi} = z^{\Lambda}\cdot e^{\Lambda\frac{\beta}{1+\beta^2}(z-\frac{1}{z})}\cdot e^{i\Lambda\chi} \tag{2.48}$$

where

$$\Lambda = \Omega\sqrt{\frac{a_o^3}{\mu}} \tag{2.49}$$

As we know, the procedure needs to expand (2.48) in powers of z (considering the orbital elements and χ constant). To make this feasible, we need to introduce another approximation, and expand the *second factor* of (2.48) - and consequently everything else - in terms of β. Even then, due to the first factor of

(2.48), the result becomes, in general, a linear combination of *non-integer* powers of z. This will then be true, in each iteration, of the $\mathcal{Q}_{(n)}(z)$ and $\mathcal{W}_{(n)}(z)$ quantities as well, thus

$$\mathcal{Q}_{(n)}(z) = \sum_{\ell} \mathcal{Q}_{\ell} z^{\delta_\ell} \tag{2.50a}$$

$$\mathcal{W}_{(n)}(z) = \sum_{\ell} \mathcal{W}_{\ell} z^{\delta_\ell} \tag{2.50b}$$

where the summation is over some discrete set of exponents, some integer, some not. Each sum can be clearly reorganized in the following manner:

$$\mathcal{Q}_{(n)}(z) = \sum_{k=-\infty}^{\infty} \left(\sum_j \mathcal{Q}_{k,j} z^{\alpha_{k,j}} \right) z^k \tag{2.51a}$$

$$\mathcal{W}_{(n)}(z) = \sum_{k=-\infty}^{\infty} \left(\sum_j \mathcal{W}_{k,j} z^{\alpha_{k,j}} \right) z^k \tag{2.51b}$$

where the j-summation usually consist of only a handful of terms and the α's are chosen from the $[-1/2, 1/2]$ range, in a 'symmetric' manner, e.g. $(-1/2, 1/2]$ for positive k and $[-1/2, 1/2)$ for negative k, to explicitly preserve the $\mathcal{W}_{(n)}(z)^* = -\mathcal{W}_{(n)}(z)$ property.

2.3.2 Results

The basic idea of finding a solution is the same as in the previous section: we substitute (1.105) into Eq. 2.6 and match coefficients of all powers of z. The only new thing is that now $a'_{(n)}$, $\beta'_{(n)}$, $s'_{\mathrm{p}(n)}$, $Z_1^{(n)}$, $Z_2^{(n)}$, $Z_3^{(n)}$ and $\chi'_{(n)}$ will acquire an *explicit* s-dependence of the $z^{\alpha_{k,j}}$ and $z^{-\alpha_{k,j}}$ type (yet, they have to stay *real*), 'inherited' from the z^i-coefficients of the $\mathcal{Q}_{(n)}(z)$ and $\mathcal{W}_{(n)}(z)$ expansions; the corresponding expressions for computing $a'_{(n)}$, $\beta'_{(n)}$, $s'_{\mathrm{p}(n)}$, $Z_1^{(n)}$, $Z_2^{(n)}$, $Z_3^{(n)}$ and $\chi'_{(n)}$ will thus have to be modified accordingly. Since, at this level, the procedure remains *linear*, we can replace the j summation in each (2.51a) and (2.51b) by a single term (i.e. $Q_k z^{\alpha_k}$ and $W_k z^{\alpha_k}$ respectively) - when we learn how to deal with one such term, we know how to deal with them all.

Since deriving the new formulas follows the steps of the previous, time-independent section, we will mention only the necessary extensions and modifications of what has already been presented there. Thus, when computing $\mathbb{U}'_{\mathrm{o}}$, we will be getting the following extra term, in addition to (2.19),

$$-2\mathfrak{k}\frac{q}{1+\beta z} z \frac{\mathrm{d}b_{(n)}}{\mathrm{d}z} \tag{2.52}$$

and the $\mathbb{U}''_{\mathrm{o}}$ expansion (2.20) needs to be extended by

$$\varepsilon^n \sqrt{\frac{a}{1+\beta^2}} (q + \beta q^{-1}) 2\mathrm{i}z \left(\frac{1}{2a} \frac{\mathrm{d}a'_{(n)}}{\mathrm{d}z} - \frac{\beta}{1+\beta^2} \frac{\mathrm{d}\beta'_{(n)}}{\mathrm{d}z} \right) + \varepsilon^n \sqrt{\frac{a}{1+\beta^2}} q^{-1} 2\mathrm{i}z \frac{\mathrm{d}\beta'_{(n)}}{\mathrm{d}z}$$
$$+ \sqrt{\frac{a}{1+\beta^2}} (q - \beta q^{-1}) 2z \frac{\mathrm{d}s'_{\mathrm{p}(n)}}{\mathrm{d}z} - 8\mathfrak{j}\frac{q}{(1+\beta z)^2} z \frac{\mathrm{d}b_{(n)}}{\mathrm{d}z} - 4\mathfrak{j}\frac{q}{1+\beta z} z^2 \frac{\mathrm{d}^2 b_{(n)}}{\mathrm{d}z^2} \tag{2.53}$$

since z differentiation no longer increases the power of ε (due to the extra z^{α_k} and $z^{-\alpha_k}$ factors). Similarly, (2.22) will now have to include

$$\mathbb{U}_{\mathrm{o}} \circ \mathbf{Z}'_{\mathrm{o}(n)} \doteqdot 2\mathrm{i}z\mathcal{U}_0 \circ \frac{\mathrm{d}\mathbf{Z}_{\mathrm{o}(n)}}{\mathrm{d}z} \tag{2.54}$$

Note that, coincidentally, $\varepsilon^n \Gamma_{(n)}$ remains the same as in (2.23).

2.3.2.1 Complex part

Due to the new terms, (2.26) has to be extended by

$$z\left(\frac{\mathrm{i}}{a}\frac{\mathrm{d}a'_{(n)}}{\mathrm{d}z}-\frac{\mathrm{d}Z_3^{(n)}}{\mathrm{d}z}\right)(1+\beta z)(1+\frac{\beta}{z})-2\mathrm{i}\frac{\mathrm{d}\beta'_{(n)}}{\mathrm{d}z}\frac{1-\beta^2z^2}{1+\beta^2}+2z\frac{\mathrm{d}s'_{\mathrm{p}(n)}}{\mathrm{d}z}(1+\beta z)(1-\frac{\beta}{z}) \tag{2.55}$$

Adding this to

$$(1+\frac{\beta}{z})(1+\beta z)^2\sum_{k=-\infty}^{\infty}(\mathcal{Q}_k z^{\alpha_k})z^k \tag{2.56}$$

must cancel out all coefficients of z^k.

This is achieved by

$$a'_{(n)}=\frac{a}{1+\beta^2}\mathrm{Im}\left[\frac{\alpha_1\beta}{(1+\alpha_1)^2}\mathcal{Q}_1z^{\alpha_1}+\left(\frac{(2+\alpha_0)\beta^2}{(1+\alpha_0)^2}+\frac{2-\alpha_0(1+\beta^2)}{1-\alpha_0^2}\right)\mathcal{Q}_0z^{\alpha_0}\right.$$
$$\left.-\beta\left(\frac{2-\alpha_{-1}}{(1-\alpha_{-1})^2}+\frac{2\beta^2+\alpha_{-1}(1+\beta^2)}{1-\alpha_{-1}^2}\right)\mathcal{Q}_{-1}z^{\alpha_{-1}}+\frac{\alpha_{-2}\beta^2}{(1-\alpha_{-2})^2}\mathcal{Q}_{-2}z^{\alpha_{-2}}\right] \tag{2.57a}$$

$$\beta'_{(n)}=-\frac{1}{4}\mathrm{Im}\left[\frac{1+\beta^2+2\alpha_1\beta^2}{(1+\alpha_1)^2}\mathcal{Q}_1z^{\alpha_1}-\left(\frac{\beta-3\beta^3-2\alpha_0\beta^3}{(1+\alpha_0)^2}-\frac{4\beta}{1-\alpha_0^2}\right)\mathcal{Q}_0z^{\alpha_0}\right.$$
$$\left.+\left(\frac{3-\beta^2-2\alpha_{-1}}{(1-\alpha_{-1})^2}+\frac{4\beta^2}{1-\alpha_{-1}^2}\right)\mathcal{Q}_{-1}z^{\alpha_{-1}}+\beta\frac{1+\beta^2-2\alpha_{-2}}{(1-\alpha_{-2})^2}\mathcal{Q}_{-2}z^{\alpha_{-2}}\right] \tag{2.57b}$$

$$Z_3^{(n)}=\frac{1}{4\beta}\mathrm{Re}\left[-\left(\frac{1-\beta^2}{(1+\alpha_1)^2}+\frac{2\beta^2}{1+\alpha_1}\right)\mathcal{Q}_1z^{\alpha_1}+\beta\left(\frac{2}{1-\alpha_0}-\frac{2(1+\beta^2)}{1+\alpha_0}+\frac{1-\beta^2}{(1+\alpha_0)^2}\right)\mathcal{Q}_0z^{\alpha_0}\right.$$
$$\left.+\left(\frac{2(1+\beta^2)}{1-\alpha_{-1}}+\frac{1-\beta^2}{(1-\alpha_{-1})^2}-\frac{2\beta^2}{1+\alpha_{-1}}\right)\mathcal{Q}_{-1}z^{\alpha_{-1}}+\beta\frac{1+\beta^2-2\alpha_{-2}}{(1-\alpha_{-2})^2}\mathcal{Q}_{-2}z^{\alpha_{-2}}\right] \tag{2.57c}$$

$$s'_{\mathrm{p}(n)}-\frac{Z_3^{(n)}}{2}=-\frac{1}{4}\mathrm{Re}\left[\beta\left(\frac{\beta^2}{1+\beta^2}\frac{1}{(1+\alpha_1)^2}-\frac{2}{1+\alpha_1}\right)\mathcal{Q}_1z^{\alpha_1}\right.$$
$$+\left(\frac{-1+2\alpha_0(1+\beta^2)}{1-\alpha_0^2}-\frac{\beta^4}{1+\beta^2}\frac{3+\alpha_0}{(1+\alpha_0)(1-\alpha_0^2)}\right)\mathcal{Q}_0z^{\alpha_0}$$
$$\left.+\beta\left(\frac{1+2\alpha_{-1}}{1-\alpha_{-1}^2}-\frac{\beta^2}{1+\beta^2}\frac{3-\alpha_{-1}}{(1-\alpha_{-1})(1-\alpha_{-1}^2)}\right)\mathcal{Q}_{-1}z^{\alpha_{-1}}+\frac{\beta^4\mathcal{Q}_{-2}z^{\alpha_{-2}}}{(1+\beta^2)(1-\alpha_{-2})^2}\right] \tag{2.57d}$$

and

$$\mathcal{D}_{(n)}(z)=-\frac{1}{4}\sum_{\substack{k=-\infty\\k\neq-1,0}}^{\infty}\left(\frac{\beta(k+\frac{1}{2}+\alpha_{k-1})\mathcal{Q}_{k-1}z^{\alpha_{k-1}}}{(k+\alpha_{k-1})^2(k+1+\alpha_{k-1})}+\frac{(k-\frac{1}{2}+\alpha_k)\mathcal{Q}_kz^{\alpha_k}}{(k+\alpha_k)^2(k+1+\alpha_k)}\right.$$
$$+\frac{\beta^2(k+\frac{3}{2}+\alpha_k)\mathcal{Q}_kz^{\alpha_k}}{(k+\alpha_k)(k+1+\alpha_k)^2}+\frac{\beta(k+\frac{1}{2}+\alpha_{k+1})\mathcal{Q}_{k+1}z^{\alpha_{k+1}}}{(k+\alpha_{k+1})(k+1+\alpha_{k+1})^2}-\frac{\frac{1}{2}\beta^2\mathcal{Q}^*_{-k-2}z^{-\alpha_{-k-2}}}{(k-\alpha_{-k-2})(k+1-\alpha_{-k-2})^2}$$
$$\left.-\frac{\frac{1}{2}\beta\mathcal{Q}^*_{-k-1}z^{-\alpha_{-k-1}}}{(k-\alpha_{-k-1})^2(k+1-\alpha_{-k-1})^2}+\frac{\frac{1}{2}\mathcal{Q}^*_{-k}z^{-\alpha_{-k}}}{(k-\alpha_{-k})^2(k+1-\alpha_{-k})}\right)z^k \tag{2.57e}$$

Note that, when all α's are equal to zero, these formulas agree with (2.29), as expected.

Proof. Using (2.57e), we first compute (again, organize your work by collecting $Q_k z^{\alpha_k}$ and $Q_k^* z^{-\alpha_k}$ proportional terms, for each integer k):

$$(1+\frac{\beta}{z})(1+\beta z)^2 \sum_{k=-\infty}^{\infty} (\mathcal{Q}_k z^{\alpha_k}) z^k - 2\mathcal{D}_{(n)}(z) - 2\mathcal{D}^*_{(n)}(z) - 2z(5+3z\beta)\frac{\mathrm{d}\mathcal{D}_{(n)}(z)}{\mathrm{d}z} + 2(z-\beta)$$
$$\frac{\mathrm{d}\mathcal{D}^*_{(n)}(z)}{\mathrm{d}z} - 4z(1+\beta z)\frac{\mathrm{d}^2\mathcal{D}_{(n)}(z)}{\mathrm{d}z^2} = \left(4\beta^2 + \frac{\beta(1+\beta^2) - 4\alpha_{-2} + 4\alpha_{-2}^2)}{z(1-\alpha_{-2}^2)}\right) Q_{-2} z^{\alpha_{-2}}$$
$$-\beta(1+\beta^2)\frac{1-2\alpha_{-2}}{z(1-\alpha_{-2}^2)} Q^*_{-2} z^{-\alpha_{-2}} + \left(4\beta(2+\beta^2) + \frac{3-\beta^2-4(2+\beta^2)\alpha_{-1}+4(1+\beta^2)\alpha_{-1}^2)}{z(1-\alpha_{-1}^2)} + 4\beta^2 z\right) Q_{-1} z^{\alpha_{-1}}$$
$$-\frac{(1+\beta^2)(3-2\alpha_{-1})}{z(1-\alpha_{-1}^2)} Q^*_{-1} z^{-\alpha_{-1}} + \left(4(1+2\beta^2) + \frac{4\beta}{z} + \beta\frac{3\beta^2-1+4(1+2\beta^2)\alpha_0 + 4(1+\beta^2)\alpha_0^2)}{(1+\alpha_0)^2} z\right) Q_0 z^{\alpha_0}$$
$$-\frac{\beta(1+\beta^2)(3+2\alpha_0)}{z(1+\alpha_0)^2} Q_0^* z^{-\alpha_0} + \left(4\beta + \frac{1+\beta^2+4\beta^2\alpha_1 + 4\beta^2\alpha_1^2)}{(1+\alpha_1)^2} z\right) Q_1 z^{\alpha_1} + \frac{(1+\beta^2)(1+2\alpha_1)}{z(1+\alpha_1)^2} Q_1^* z^{-\alpha_1} \tag{2.58}$$

Based on the autonomous solution, we assume that $a'_{(n)}$ and $\beta'_{(n)}$ can each be expressed as a linear combination of $\mathrm{Im}(Q_\ell z^{\alpha_\ell})$, and $Z_3^{(n)}$ and $s'_{\mathrm{p}(n)}$ similarly as a linear combination of $\mathrm{Re}(Q_\ell z^{\alpha_\ell})$, where ℓ goes from -2 to 1. Note that

$$z\frac{\mathrm{d}\,\mathrm{Im}(Q_\ell z^{\alpha_\ell})}{\mathrm{d}z} = -\mathrm{i}\alpha_\ell \,\mathrm{Re}(Q_\ell z^{\alpha_\ell}) \tag{2.59a}$$
$$z\frac{\mathrm{d}\,\mathrm{Re}(Q_\ell z^{\alpha_\ell})}{\mathrm{d}z} = \mathrm{i}\alpha_\ell \,\mathrm{Im}(Q_\ell z^{\alpha_\ell}) \tag{2.59b}$$

(this is why the corresponding system of equations no longer de-couples). Making

$$\mathrm{i}\left(1-\beta^2-\beta z+\frac{\beta}{z}\right)\frac{a'_{(n)}}{2a} - \mathrm{i}(z+\frac{1}{z})\beta'_{(n)} - 4\mathrm{i}\frac{\beta}{1+\beta^2}\beta'_{(n)} - \left(2(1-\beta^2)+\beta z - \frac{\beta}{z}\right) Z_3^{(n)} + 4(1+\beta^2)s'_{\mathrm{p}(n)}$$
$$+2\mathrm{i}z\frac{\mathrm{d}\beta'_{(n)}}{\mathrm{d}z}\frac{\frac{1}{z}-\beta z}{1+\beta^2} + z\left(\frac{\mathrm{d}Z_3^{(n)}}{\mathrm{d}z} - \frac{\mathrm{i}}{a}\frac{\mathrm{d}a'_{(n)}}{\mathrm{d}z}\right)\left(1+\beta^2+\beta z+\frac{\beta}{z}\right) + 2z\frac{\mathrm{d}s'_{\mathrm{p}(n)}}{\mathrm{d}z}\left(1-\beta^2+\beta z - \frac{\beta}{z}\right) \tag{2.60}$$

equal to (2.58), individually, for each coefficient of $\mathrm{Re}(Q_\ell z^{\alpha_\ell})$ and $\mathrm{Im}(Q_\ell z^{\alpha_\ell})$ of the z^1, z^0 and z^{-1}-proportional part of the expression, results in 24 linear equations for the 16 unknown coefficients. Even though overdetermined, this set can be solved, yielding (2.57), as can be easily verified (e.g. solve the $\mathrm{Re}(Q_\ell z^{\alpha_\ell})$-related set of 12 equations plus any three of the remaining equations; the solution is unique and independent of which three equations we select). □

2.3.2.2 j-𝔨 part

Now, more briefly: (2.37) acquires the following new terms

$$\mathrm{i}(1-\beta^2)(z-\frac{1}{z})z\frac{\mathrm{d}Z_1^{(n)}}{\mathrm{d}z} + \left((1+\beta^2)(z+\frac{1}{z})+4\beta\right) z\frac{\mathrm{d}Z_2^{(n)}}{\mathrm{d}z} + 8\mathrm{i}\left(1-\frac{\beta(z-\frac{1}{z})}{(1+\beta z)(1+\frac{\beta}{z})}\right) z\frac{\mathrm{d}b_{(n)}}{\mathrm{d}z} + 8\mathrm{i}z^2\frac{\mathrm{d}^2 b_{(n)}}{\mathrm{d}z^2} \tag{2.61}$$

Together with

$$(1+\frac{\beta}{z})(1+\beta z)\left(\sum_{k=0}^{\infty}(\mathcal{W}_k z^{\alpha_k})z^k - \sum_{k=0}^{\infty}(\mathcal{W}_k^* z^{-\alpha_k})z^{-k}\right) \tag{2.62}$$

the sum of the two expressions must be identically equal to zero. This is achieved by

$$Z_1^{(n)} = -\frac{1}{1-\beta^2}\mathrm{Im}\left[\frac{2\beta}{1-\alpha_0^2}\mathcal{W}_0 z^{\alpha_0} + \frac{1+\beta^2}{2+\alpha_1}\mathcal{W}_1 z^{\alpha_1} + \frac{\beta\alpha_2}{(1+\alpha_2)(2+\alpha_2)}\mathcal{W}_2 z^{\alpha_2}\right] \tag{2.63a}$$

$$Z_2^{(n)} = \frac{1}{1+\beta^2}\mathrm{Re}\left[\frac{2\beta\alpha_0}{1-\alpha_0^2}\mathcal{W}_0 z^{\alpha_0} - \frac{1+\beta^2}{2+\alpha_1}\mathcal{W}_1 z^{\alpha_1} - \frac{\beta\alpha_2}{(1+\alpha_2)(2+\alpha_2)}\mathcal{W}_2 z^{\alpha_2}\right] \tag{2.63b}$$

$$b_{(n)} = \frac{1}{4(1+\beta^2)}\mathrm{Im}\left[\frac{(1-\beta^2)^2}{1-\alpha_0^2}\mathcal{W}_0 z^{\alpha_0} + \frac{(1+\beta^2)\beta}{(1+\alpha_1)(2+\alpha_1)}\mathcal{W}_1 z^{\alpha_1} + \frac{2\beta^2}{(1+\alpha_2)(2+\alpha_2)}\mathcal{W}_2 z^{\alpha_2}\right] \tag{2.63c}$$

$$\mathcal{S}_{(n)}(z) = -\frac{\mathrm{i}}{4}\mathrm{Im}\left[\sum_{k=2}^{\infty}\left(\frac{\beta\mathcal{W}_{k-1}z^{\alpha_{k-1}}}{(k+\alpha_{k-1}-1)(k+\alpha_{k-1})} + \frac{(1+\beta^2)\mathcal{W}_k z^{\alpha_k}}{(k+\alpha_k)^2-1} + \frac{\beta\mathcal{W}_{k+1}z^{\alpha_{k+1}}}{(k+\alpha_{k+1})(k+\alpha_{k+1}+1)}\right)z^k\right] \tag{2.63d}$$

Again, these formulas reduce to (2.40), but one has to be careful about $\mathrm{Im}[\mathcal{W}_0 z^{\alpha_0}]$ becoming, in the $\alpha_0 \to 0$ limit, only one half of the old $\mathrm{Im}[\mathcal{W}_0]$ - the other half coming from the same limit of $\mathrm{Im}[\mathcal{W}_0^* z^{-\alpha_0}]$.

Proof. First, we compute

$$\begin{aligned}
&(1+\beta z)(1+\frac{\beta}{z})\mathcal{W}_{(n)}(z) - 8\frac{1+\beta^2}{(1+\beta z)(1+\frac{\beta}{z})}\mathcal{S}_{(n)}(z) + 8\frac{z+2\beta+z\beta^2}{(1+\beta z)(1+\frac{\beta}{z})}\frac{\mathrm{d}\mathcal{S}_{(n)}(z)}{\mathrm{d}z} + 8z^2\frac{\mathrm{d}^2\mathcal{S}_{(n)}(z)}{\mathrm{d}z^2} \\
&= (1+\beta z)(1+\frac{\beta}{z})(\mathcal{W}_0 z^{\alpha_0} - \mathcal{W}_0^* z^{-\alpha_0}) \\
&+ \frac{\frac{\beta^2}{z} + 2\beta(1+\beta^2) + (1+\frac{1+3\alpha_1}{1+\alpha_1}\beta^2+\beta^4)z + \frac{1+\alpha_1}{2+\alpha_1}\beta(1+\beta^2)z^2}{(1+\beta z)(1+\frac{\beta}{z})}\mathcal{W}_1 z^{\alpha_1} \\
&- \frac{\beta^2 z + 2\beta(1+\beta^2) + (1+\frac{1+3\alpha_1}{1+\alpha_1}\beta^2+\beta^4)\frac{1}{z} + \frac{1+\alpha_1}{2+\alpha_1}\beta(1+\beta^2)\frac{1}{z^2}}{(1+\beta z)(1+\frac{\beta}{z})}\mathcal{W}_1^* z^{-\alpha_1} \\
&+ \beta\frac{\frac{\beta}{z} + \frac{\alpha_2}{1+\alpha_2}(1+\beta^2) + \frac{\alpha_2}{2+\alpha_2}\beta z}{(1+\beta z)(1+\frac{\beta}{z})}\mathcal{W}_2 z^{\alpha_2} - \beta\frac{\beta z + \frac{\alpha_2}{1+\alpha_2}(1+\beta^2) + \frac{\alpha_2}{2+\alpha_2}\beta\frac{1}{z}}{(1+\beta z)(1+\frac{\beta}{z})}\mathcal{W}_2^* z^{-\alpha_2}
\end{aligned} \tag{2.64}$$

This, added to

$$\begin{aligned}
&\mathrm{i}\frac{(1-\beta^2)\left(2\beta+z+\frac{1}{z}\right)\left(2+\beta(z+\frac{1}{z})\right)}{(1+\beta z)(1+\frac{\beta}{z})}Z_1^{(n)} + \mathrm{i}(1-\beta^2)(z-\frac{1}{z})z\frac{\mathrm{d}Z_1^{(n)}}{\mathrm{d}z} \\
&+ \frac{(z-\frac{1}{z})\left(2+2\beta^4+\beta(1+\beta^2)(z+\frac{1}{z})\right)}{(1+\beta z)(1+\frac{\beta}{z})}Z_2^{(n)} + \left((1+\beta^2)(z+\frac{1}{z})+4\beta\right)z\frac{\mathrm{d}Z_2^{(n)}}{\mathrm{d}z} \\
&-8\mathrm{i}\frac{1+\beta^2}{(1+\beta z)(1+\frac{\beta}{z})}b_{(n)} + 8\mathrm{i}\left(1-\frac{\beta(z-\frac{1}{z})}{(1+\beta z)(1+\frac{\beta}{z})}\right)z\frac{\mathrm{d}b_{(n)}}{\mathrm{d}z} + 8\mathrm{i}z^2\frac{\mathrm{d}^2 b_{(n)}}{\mathrm{d}z^2}
\end{aligned} \tag{2.65}$$

and multiplied by $(1+\beta z)(1+\beta/z)$, should be identically equal to zero. We assume that $Z_1^{(n)}$ and $b_{(n)}$ can each be expressed as a linear combination of $\mathrm{Im}(W_\ell z^{\alpha_\ell})$, and $Z_2^{(n)}$ similarly as a linear combination of $\mathrm{Re}(W_\ell z^{\alpha_\ell})$, where ℓ goes from 0 to 2. Note that, in addition to the $Q \to W$ analog of (2.59a), we also have

$$z\frac{\mathrm{d}}{\mathrm{d}z}\left(z\,\frac{\mathrm{d}\,\mathrm{Im}(W_\ell z^{\alpha_\ell})}{\mathrm{d}z}\right) = \alpha_\ell^2\,\mathrm{Im}(W_\ell z^{\alpha_\ell}) \tag{2.66}$$

$$z\frac{\mathrm{d}}{\mathrm{d}z}\left(z\,\frac{\mathrm{d}\,\mathrm{Re}(W_\ell z^{\alpha_\ell})}{\mathrm{d}z}\right) = \alpha_\ell^2\,\mathrm{Re}(W_\ell z^{\alpha_\ell}) \tag{2.67}$$

Collecting z^2-proportional terms and making each $\mathrm{Im}(W_\ell z^{\alpha_\ell})$ and $\mathrm{Re}(W_\ell z^{\alpha_\ell})$ component equal to zero yields six linear equations for the six coefficients of $Z_1^{(n)}$ and $Z_2^{(n)}$; these have a unique solution (2.63a) and (2.63b). Similarly collecting z^1-proportional terms yields two sets of three equations (by collecting the $\mathrm{Im}(W_\ell z^{\alpha_\ell})$ and the $\mathrm{Re}(W_\ell z^{\alpha_\ell})$ components) for the three coefficients of $b_{(n)}$; luckily, they have identical solutions, given in (2.63c). The same solution also eliminates the three $\mathrm{Im}(W_\ell z^{\alpha_\ell})$ components of z^0-proportional terms (the $\mathrm{Re}(W_\ell z^{\alpha_\ell})$ components are identically equal to zero). Due to the $\mathcal{W}_{(n)}^*(z) = -\mathcal{W}_{(n)}(z)$ property, the z^{-1} and z^{-2} need not be considered separately; they would lead to equations which have already been solved. □

2.4 SUMMARY

To solve (2.6) which, together with (2.5), (2.1), (1.16) and (1.28) is equivalent to (1.27), we proceed as follows:

1. Starting with

$$\mathbb{U}_{\mathrm{o}} = \sqrt{\frac{a}{1+\beta^2}}\; q \circ \left(1+\frac{\beta}{z}\right) \equiv \mathcal{U}_0 \tag{2.68}$$

   and $a' = \beta' = \mathbf{Z}_{\mathrm{o}} = s_{\mathrm{p}}' = b = \mathcal{D}(z) = \mathcal{S}(z) = \Gamma = 0$

   (a) evaluate the left hand side of (2.6),calling if $\mathbb{F}_{\mathrm{o}}^{(1)}$;

   (b) using (7.38) and (2.9), convert $\mathbb{F}_{\mathrm{o}}^{(1)}$ into two complex quantities $\mathcal{Q}_{(1)}(z)$ and $\mathcal{W}_{(1)}(z)$;

   (c) in the autonomous case ($\varepsilon\mathbf{f}$ has no explicit t dependence), update a', β' , $\mathbf{Z}_{\mathrm{o}}$, s_{p}', $b, \mathcal{D}(z)$ and $\mathcal{S}(z)$ by adding (2.29) and (2.40) respectively; when $\varepsilon\mathbf{f}$ depends, explicitly, on t, we first need (2.44) and (2.45) to express t and its s-derivatives in terms of z, and then switch to (2.57) and (2.63) instead.

2. Using

$$\mathbb{U}_{\mathrm{o}} = \sqrt{\frac{a}{1+\beta^2}}\; q \circ \left(1+\frac{\beta}{z}+\mathcal{D}(z)+\mathfrak{k}\circ\frac{ib+\mathcal{S}(z)}{1+\beta z}\right) \tag{2.69}$$

   and the updated values of a', β' , $\mathbf{Z}_{\mathrm{o}}$, s_{p}', b, $\mathcal{D}(z)$ and $\mathcal{S}(z)$, repeat the procedure with $n=2$, $n=3$, etc., to reach the desired accuracy.

3. By dividing a', β' and each component of $\mathbf{Z}_{\mathrm{o}}$ by $2(1-s_{\mathrm{p}}')$, convert s-derivatives to ω-derivatives, and expand each answer in powers of ε. At the same time, $\mathbf{Z}_{\mathrm{o}}$ is converted to ϕ', θ' and ψ' by (1.19).

4. Solve the corresponding set of first-order differential equations for a, β, ϕ, θ and ψ (with ω as the independent variable) to the same power-of-ε accuracy, by any traditional technique.

5. Using (1.17), (2.1) and (1.28a), convert the answer to the satellite's location $\mathbf{r}$ (as a function of eccentric anomaly ω). Divide (1.28b) by $2(1 - s'_{\mathrm{p}})$ and integrate the right hand side, to find the relationship between t and ω.

Even though rather non-trivial, the technique is capable of finding the solution to Kepler's problem, analytically, to an arbitrary accuracy. Furthermore, it is not plagued by the usual difficulties of other iterative techniques, such as encountering zero divisors and having to resort to the so-called AVERAGING PRINCIPLE (an euphemism for an approximation which, however accurate, makes it pointless to continue with another iteration; the approximation eliminates all oscillating terms from the rate of change of each orbital element, thus making the corresponding differential equations autonomous). Another of its beneficial features is the clear separation of orbital elements from parallel and perpendicular distortions of a single orbit (not doing this is a big handicap of OSCULATING orbital elements), which enables us to maintain the autonomous property of the resulting differential equations (for autonomous perturbing forces). And, also worth mentioning, is the ease of treatment of resonant perturbing forces (in the non-autonomous case).

At the same time, it is obvious that constructing a truly accurate solution takes an enormous amount of symbolic computation, which cannot be achieved by hand (unless one is willing to spend decades on a single problem, as was often the case until the advent of computers). Since, in the remainder of this book, we want to demonstrate the technique by means of step-by-step examples, we will be forced to present (with one exception) only ε-accurate solutions (we may thus drop the corresponding subscript from our notation, using $\mathcal{Q}(z)$ instead of $\mathcal{Q}_{(1)}(z)$, etc.). Worse yet, there will be occasions when, for the sake of simplicity, we must resort to using the averaging 'principle' as well, even though this goes against the spirit of the technique. The reader should keep in mind that, to take a full advantage of it, the technique should be converted into a (relatively simple) computer program - something we cannot incorporate into a book of this size and (modest) ambitions.

CHAPTER 3

Perturbing forces

Abstract

In this chapter we derive expressions for common perturbing forces. Note that these are defined in terms of (1.27), i.e. *per unit* of satellite's mass.

3.0.1 Terminology and notation

First we recall our terminology: the celestial object whose motion we study is called the PERTURBED body (or SATELLITE); it is orbiting the so-called PRIMARY. All other bodies (if any) are called PERTURBING. For quantities relating to the perturbed body we consistently use *small letters*, for those relating to the primary we use *capital letters*, while for the PERTURBING BODY we use either-type letters (usually small) with *subscripts*.

3.1 PLANETS OF THE SOLAR SYSTEM

When investigation the motion of a single planet, the main perturbing effects arise from the gravitational pull of the remaining planets, which are considered to be point-like objects. Using our notational scheme:

1. M, m, and m_i represent the mass of Sun, the perturbed planet, and one of the other perturbing planets, respectively, each already multiplied by the GRAVITATIONAL CONSTANT G (we will call this the *gravitational* mass, with dimensions of meter^3 sec^{-2}),
2. $\mathbf{X}$, $\mathbf{x}$ and $\mathbf{x}_i$ are the corresponding locations, in an inertial frame,
3. $\mathbf{r} \equiv \mathbf{x} - \mathbf{X}$ and $\mathbf{r}_i \equiv \mathbf{x}_i - \mathbf{X}$ are locations of the perturbed and perturbing planets with respect to the center of Sun, respectively.

By Newton's gravitational law

$$\ddot{\mathbf{x}} = -M\frac{\mathbf{r}}{r^3} - \sum_i m_i \frac{\mathbf{r} - \mathbf{r}_i}{|\mathbf{r} - \mathbf{r}_i|^3} \tag{3.1}$$

and

$$\ddot{\mathbf{X}} = m\frac{\mathbf{r}}{r^3} + \sum_i m_i \frac{\mathbf{m}_i}{r_i^3} \tag{3.2}$$

from which it follows, by simple subtraction, that

$$\begin{aligned}\ddot{\mathbf{r}} &= -(M+m)\frac{\mathbf{r}}{r^3} - \sum_i m_i \left(\frac{\mathbf{r} - \mathbf{r}_i}{|\mathbf{r} - \mathbf{r}_i|^3} + \frac{\mathbf{r}_i}{r_i^3}\right) \\ &\equiv -\mu\frac{\mathbf{r}}{r^3} - \mu\sum_i \varepsilon_i \left(\frac{\mathbf{r} - \mathbf{r}_i}{|\mathbf{r} - \mathbf{r}_i|^3} + \frac{\mathbf{r}_i}{r_i^3}\right)\end{aligned} \tag{3.3}$$

This is a Kepler problem with not only one, but *several* perturbing terms. We should realize that the iterative procedure of the last chapter works as easily with several ε_i's, as it does with one (higher-order iterations will involve mixed terms, proportional to $\varepsilon_1\varepsilon_2$ etc.).

Jan Vrbik

3.1.1 Earth-Moon correction

When dealing with the solar system, it is convenient to consider each planet and its moons as a single body. This works fairly well (as an accurate approximation) in most cases, except for the Earth-Moon pair, due to Moon's relatively large size. We can deal with this extra complication as follows:

Let $\mathbf{x}$ be the location of the Earth-Moon *center of mass*, i.e.

$$\mathbf{x} = \frac{m_\mathrm{e}}{m_\mathrm{e}+m_\mathrm{m}}\mathbf{x}_\mathrm{e} + \frac{m_\mathrm{m}}{m_\mathrm{e}+m_\mathrm{m}}\mathbf{x}_\mathrm{m} = \mathbf{x}_\mathrm{e} + \frac{m_\mathrm{m}}{m_\mathrm{e}+m_\mathrm{m}}\rho = \mathbf{x}_\mathrm{e} - \frac{m_\mathrm{e}}{m_\mathrm{e}+m_\mathrm{m}}\rho \tag{3.4}$$

where Earth and Moon-related quantities are subscribed e and m respectively, and $\rho \equiv \mathbf{x}_\mathrm{m} - \mathbf{x}_\mathrm{e}$.

Then, based on the same gravitational law

$$\begin{aligned}\ddot{\mathbf{x}} &= \frac{m_\mathrm{e}}{m_\mathrm{e}+m_\mathrm{m}}\left(-M\frac{\mathbf{x}_\mathrm{e}-\mathbf{X}}{|\mathbf{x}_\mathrm{e}-\mathbf{X}|^3} - m_\mathrm{m}\frac{\mathbf{x}_\mathrm{e}-\mathbf{x}_\mathrm{m}}{|\mathbf{x}_\mathrm{e}-\mathbf{x}_\mathrm{m}|^3}\right) + \frac{m_\mathrm{m}}{m_\mathrm{e}+m_\mathrm{m}}\left(-M\frac{\mathbf{x}_\mathrm{m}-\mathbf{X}}{|\mathbf{x}_\mathrm{m}-\mathbf{X}|^3} - m_\mathrm{e}\frac{\mathbf{x}_\mathrm{m}-\mathbf{x}_\mathrm{e}}{|\mathbf{x}_\mathrm{m}-\mathbf{x}_\mathrm{e}|^3}\right)\\ &= -M\left(\frac{m_\mathrm{e}}{m_\mathrm{e}+m_\mathrm{m}}\frac{\mathbf{r}-\frac{m_\mathrm{m}}{m_\mathrm{e}+m_\mathrm{m}}\rho}{\left|\mathbf{r}-\frac{m_\mathrm{m}}{m_\mathrm{e}+m_\mathrm{m}}\rho\right|^3} + \frac{m_\mathrm{m}}{m_\mathrm{e}+m_\mathrm{m}}\frac{\mathbf{r}+\frac{m_\mathrm{e}}{m_\mathrm{e}+m_\mathrm{m}}\rho}{\left|\mathbf{r}+\frac{m_\mathrm{e}}{m_\mathrm{e}+m_\mathrm{m}}\rho\right|^3}\right)\end{aligned} \tag{3.5}$$

where $\mathbf{r} \equiv \mathbf{x} - \mathbf{X}$.

The last expression can be expanded in terms of ρ (small, relative to $\mathbf{r}$) yielding, to the second-order (in ρ) accuracy

$$-M\left[\frac{\mathbf{r}}{r^3} - \frac{m_\mathrm{e}m_\mathrm{m}}{(m_\mathrm{e}+m_\mathrm{m})^2}\left(\frac{3}{2}\frac{\rho^2}{r^5}\mathbf{r} - \frac{15}{2}\frac{(\mathbf{r}\cdot\rho)^2}{r^7}\mathbf{r} + 3\frac{\mathbf{r}\cdot\rho}{r^5}\rho\right) + \cdots\right] \tag{3.6}$$

Proof.

$$\begin{aligned}\frac{\mathbf{r}+\boldsymbol{\Delta}}{|\mathbf{r}+\boldsymbol{\Delta}|^3} &= \frac{\mathbf{r}+\boldsymbol{\Delta}}{r^3}\left(1+\frac{2\mathbf{r}\cdot\boldsymbol{\Delta}}{r^2}+\frac{\Delta^2}{r^2}\right)^{-3/2}\\ &\simeq \frac{\mathbf{r}}{r^3}\left(1-\frac{3\mathbf{r}\cdot\boldsymbol{\Delta}}{r^2}-\frac{3}{2}\frac{\Delta^2}{r^2}+\frac{15}{2}\frac{(\mathbf{r}\cdot\boldsymbol{\Delta})^2}{r^4}+\cdots\right)+\frac{\boldsymbol{\Delta}}{r^3}\left(1-\frac{3\mathbf{r}\cdot\boldsymbol{\Delta}}{r^2}+\cdots\right)\end{aligned} \tag{3.7}$$

When this formula is applied to (3.5), terms linear in ρ cancel out, whereas the quadratic terms get multiplied by

$$\frac{m_\mathrm{e}}{m_\mathrm{e}+m_\mathrm{m}}\left(\frac{m_\mathrm{m}}{m_\mathrm{e}+m_\mathrm{m}}\right)^2 + \frac{m_\mathrm{m}}{m_\mathrm{e}+m_\mathrm{m}}\left(\frac{m_\mathrm{e}}{m_\mathrm{e}+m_\mathrm{m}}\right)^2 = \frac{m_\mathrm{e}m_\mathrm{m}}{(m_\mathrm{e}+m_\mathrm{m})^2} \tag{3.8}$$

□

Note that this time there is no perturbing body as such - it is the dumbbell shape of the Earth-Moon system which causes the perturbation of its center-of-mass motion.

One can further simplify (3.6) by assuming that the motion of Moon around Earth is uniform and circular, and confined to the plane of the ecliptic. The average value (over many of Moon's orbits) of the perturbing force is then

$$-\frac{3}{4}\frac{Mm_\mathrm{e}m_\mathrm{m}}{(m_\mathrm{e}+m_\mathrm{m})^2}\frac{\rho^2}{r^5}\mathbf{r} \tag{3.9}$$

since

$$\overline{(\mathbf{r}\cdot\rho)^2} = \frac{\rho^2}{2}r^2 \tag{3.10a}$$

$$\overline{(\mathbf{r}\cdot\rho)\rho} = \frac{\rho^2}{2}\mathbf{r} \tag{3.10b}$$

the bar denoting ρ-averaging, performed by

$$\frac{\int_0^{2\pi} \cdots d\varpi}{2\pi} \tag{3.11}$$

with $\rho_x \simeq \rho \cos \varpi$ and $\rho_y \simeq \rho \sin \varpi$.

Note that

$$\frac{3}{4} \frac{m_e m_m}{(m_e + m_m)^2} \frac{\rho^2}{r^2} \tag{3.12}$$

(the *relative magnitude* of the perturbing force) is equal to 0.000000059; its small value amply justifies the approximations made.

3.2 MOON PERTURBED BY SUN

Moon is now the perturbed body of (gravitational) mass m, Earth is the primary of mass M, and the perturbing force is provided by Sun of mass M_s (the respective locations are labeled accordingly).

As usually we take $\mathbf{r} = \mathbf{x} - \mathbf{X}$, but it is more convenient now to define $\mathbf{R}_s$ as $\mathbf{X}_s - \mathbf{X}_c$, with $\mathbf{X}_c$ being the Earth-Moon *center of mass* of the previous section, namely

$$\begin{aligned} \mathbf{X}_c &= \frac{M}{M+m}\mathbf{X} + \frac{m}{M+m}\mathbf{x} \\ &= \mathbf{X} + \frac{m}{M+m}\mathbf{r} \end{aligned} \tag{3.13}$$

whose motion is described by

$$\ddot{\mathbf{X}}_c = \frac{MM_s}{M+m} \frac{\mathbf{R}_s + \frac{m}{M+m}\mathbf{r}}{|\mathbf{R}_s + \frac{m}{M+m}\mathbf{r}|^3} + \frac{mM_s}{M+m} \frac{\mathbf{R}_s - \frac{M}{M+m}\mathbf{r}}{|\mathbf{R}_s - \frac{M}{M+m}\mathbf{r}|^3} \tag{3.14}$$

identical to (3.5), except for a new notation ($\mathbf{x} \to \mathbf{X}_c$, $\rho \to \mathbf{r}$, $\mathbf{r} \to -\mathbf{R}_s$, $m_m \to m$, $m_e \to M$ and $M \to M_s$), to comply with our convention.

Similarly,

$$\begin{aligned} \ddot{\mathbf{r}} &= \left(-M_s \frac{\mathbf{x} - \mathbf{X}_s}{|\mathbf{x} - \mathbf{X}_s|^3} - M \frac{\mathbf{x} - \mathbf{X}}{|\mathbf{x} - \mathbf{X}|^3}\right) - \left(-M_s \frac{\mathbf{X}_s - \mathbf{X}}{|\mathbf{X}_s - \mathbf{X}|^3} - m \frac{\mathbf{X} - \mathbf{x}}{|\mathbf{X} - \mathbf{x}|^3}\right) \\ &= -(M+m)\frac{\mathbf{r}}{r^3} - M_s \left(\frac{\mathbf{R}_s + \frac{m}{M+m}\mathbf{r}}{|\mathbf{R}_s + \frac{m}{M+m}\mathbf{r}|^3} - \frac{\mathbf{R}_s - \frac{M}{M+m}\mathbf{r}}{|\mathbf{R}_s - \frac{M}{M+m}\mathbf{r}|^3}\right) \end{aligned} \tag{3.15}$$

since

$$\mathbf{X}_s - \mathbf{X} = \frac{m}{M+m}\mathbf{r} + \mathbf{R}_s \tag{3.16a}$$

$$\mathbf{x} - \mathbf{X}_s = \frac{M}{M+m}\mathbf{r} - \mathbf{R}_s \tag{3.16b}$$

This time, it is $\mathbf{r}$ whose magnitude is small relative to $\mathbf{R}_s$, implying that the perturbing force can be expanded as follows:

$$-\frac{M_s}{R_s^3}\left(\mathbf{r} - 3(\mathbf{r}\cdot\mathbf{R}_s)\frac{\mathbf{R}_s}{R_s^2}\right) - \frac{M_s}{R_s^5}\frac{M-m}{M+m}\left(\frac{3}{2}r^2\mathbf{R}_s - \frac{15}{2}(\mathbf{r}\cdot\mathbf{R}_s)^2\frac{\mathbf{R}_s}{R_s^2} + 3(\mathbf{r}\cdot\mathbf{R}_s)\mathbf{r}\right) + \ldots \tag{3.17}$$

To help us assess the relative importance of each of the two perturbing terms, we note that

$$\frac{M}{M+m}\frac{r^3}{R_s^3} \simeq 0.00558 \tag{3.18a}$$

$$\frac{M}{M+m}\frac{M-m}{M+m}\frac{r^4}{R_s^4} \simeq 0.000014 \tag{3.18b}$$

the second term being smaller than the second power of the first term.

3.3 ZONAL HARMONICS

For satellites orbiting near Earth, the point-mass approximation is no longer adequate, and Earth's exact shape (different from a perfect sphere) introduces yet another important set of perturbations. The following is a quick review of relevant formulas.

The gravitation potential of any *axi-symmetric* body (Earth, in our case) is given by

$$V(\mathbf{r}) = -\int_0^{2\pi}\int_0^{\pi}\int_0^{R(\tilde\theta)} \frac{\rho(\tilde r, \tilde\theta)}{|\mathbf{r}-\tilde{\mathbf{r}}|}\tilde r^2 \sin\tilde\theta \; d\tilde r \; d\tilde\theta \; d\tilde\varphi \tag{3.19}$$

where $\rho(\tilde r, \tilde\theta)$ is the body's gravitational-mass density ($\tilde\varphi$-independent, as we have chosen z to agree with the symmetry axis) and $R(\tilde\theta)$ is its $\tilde\theta$-dependent 'radius' (surface distance from center).

Using the well known expansion [21]

$$\frac{1}{|\mathbf{r}-\tilde{\mathbf{r}}|} = \frac{1}{r}\sum_{n=0}^{\infty}\left(\frac{\tilde r}{r}\right)^n P_n(\cos\gamma) \tag{3.20}$$

(where P_n are Legendre polynomials and γ is the angle between the directions of $\mathbf{r}$ and $\tilde{\mathbf{r}}$), and the addition formula for spherical harmonics [1]

$$P_n(\cos\gamma) \equiv \sum_{m=0}^{n}(2-\delta_{0m})\frac{(n-m)!}{(n+m)!}P_n^m(\cos\tilde\theta)P_n^m(\cos\theta)\cos[m(\tilde\varphi-\varphi)] \tag{3.21}$$

(where P_n^m are *associate* Legendre polynomials) one can integrate (3.19) (note that, for axi-symmetric body, the $m \neq 0$ terms integrate to zero) to obtain

$$V(\mathbf{r}) = -\frac{M}{r}\left[1-\sum_{n=2}^{\infty}J_n\left(\frac{R_\ominus}{r}\right)^n P_n(\cos\theta)\right] \tag{3.22}$$

where M is the body's gravitational mass, $R_\ominus$ is its equatorial radius, and

$$J_n \equiv -\frac{\int_0^{\pi}\int_0^{R(\theta)} \tilde r^n P_n(\cos\tilde\theta)\rho(\tilde r,\tilde\theta)\tilde r^2\sin\tilde\theta \; d\tilde r \; d\tilde\theta}{R_\ominus^n \int_0^{\pi}\int_0^{R(\theta)} \rho(\tilde r,\tilde\theta)\tilde r^2 \sin\tilde\theta \; d\tilde r \; d\tilde\theta} \tag{3.23}$$

Trivially, $J_0 \equiv -1$ and $J_1 \equiv 0$, assuming that the origin is placed at Earth's center of mass - this explains the disappearance of the $n = 1$ term in (3.22). Empirically (for Earth) $J_2 \backsimeq 1.083 \times 10^{-3}$, $J_3 \backsimeq -2.5 \times 10^{-6}$, $J_4 \backsimeq -1.6 \times 10^{-6}$, etc. The most important of these is clearly J_2, expressing the degree of Earth's flattening or OBLATENESS (due to Earth's rotation). Therefore, only J_2 and the corresponding perturbations are considered in detail in this book, implying that of all Legendre polynomials we need only

$$P_2(x) = \frac{3x^2 - 1}{2} \tag{3.24}$$

When we discard terms beyond $n = 2$, (3.22) reads

$$V(\mathbf{r}) = -\frac{M}{r} + \frac{J_2 M R_\ominus^2}{2r^3}(3\cos^2\theta - 1) \tag{3.25}$$

where the first term is the usual, perfect-sphere potential, and the second term leads to the oblateness perturbation, given by its negative gradient, namely

$$\varepsilon\mathbf{f} = -J_2 M R_\ominus^2 \nabla \frac{3z^2 - r^2}{2r^5} = -\frac{J_2 M R_\ominus^2}{r^5}\left(\frac{3}{2}\mathbf{r} - \frac{15}{2}\frac{(\mathbf{r}\cdot\mathbf{u})^2\mathbf{r}}{r^2} + 3(\mathbf{r}\cdot\mathbf{u})\mathbf{u}\right) \tag{3.26}$$

where $\mathbf{u}$ is the unit direction of the symmetry axis (considered fixed). Note that the relative magnitude (compared to $M\mathbf{r}/r^3$) of this perturbation is proportional not only to J_2, but also to $(R_\ominus/r)^2$, thus quickly diminishing with distance from Earth.

3.3.0.1 Earth's oblateness

This section explains Earth's flattening at the poles.

Earth, consisting of an essentially elastic mass, will adjust its shape to achieve a constant value of the overall (gravitational plus rotational) potential, given to a good approximation - take the first two non-zero terms of (3.22) - by

$$-\frac{M}{R} + \frac{M J_2 R_\ominus^2 (3\cos^2\theta - 1)}{2R^3} - \frac{1}{2}\Omega^2 R^2 \sin^2\theta \tag{3.27}$$

where R is Earth's 'radius' (a function of θ). The last term is based on the rotational potential

$$-\frac{1}{2}\Omega^2\left(R^2 - (\mathbf{u}\cdot\mathbf{R})^2\right) \tag{3.28}$$

Replacing R by $R_\ominus + \Delta$, and assuming that Δ, J_2 and Ω^2 are small, we can rewrite this as

$$-\frac{M}{R_\ominus}\left(1 - \frac{\Delta}{R_\ominus} + \cdots\right) + \frac{M J_2(3\cos^2\theta - 1)}{2R_\ominus} - \frac{1}{2}\Omega^2 R_\ominus^2 \sin^2\theta \tag{3.29}$$

To find Δ as a function of θ, we make this expression equal to a constant, whose value must be $-M/R_\ominus - MJ_2/(2R_\ominus) - 1/2 \cdot \Omega^2 R_\ominus^2$, to make $\Delta = 0$ at $\theta = \pi/2$. We thus get

$$\frac{M\Delta}{R_\ominus^2} + \frac{3MJ_2\cos^2\theta}{2R_\ominus} + \frac{1}{2}\Omega^2 R_\ominus^2 \cos^2\theta = 0 \tag{3.30}$$

from which it follows that

$$\Delta \simeq -\left(\frac{3}{2}R_\ominus J_2 + \frac{\Omega^2 R_\ominus^4}{2M_0}\right)\cos^2\theta = (R_\odot - R_\ominus)\cos^2\theta \tag{3.31}$$

where $R_\odot$ is Earth's polar radius.

To complete the exercise, we show that a body of uniform mass density, flattened in the manner of the previous equation, has

$$J_2 \simeq \frac{2}{5}\left(1 - \frac{R_\odot}{R_\ominus}\right) \tag{3.32}$$

Note that, since planets normally have a heavy core and a lighter crust, J_2 is usually smaller than this, e.g. for Earth $J_2 \simeq 0.33(1 - R_\odot/R_\ominus)$.

Proof. Based on (3.23) with a constant mass density (which then cancels out) we get

$$J_2 = -\frac{\int_0^\pi \int_0^{R(\tilde\theta)} P_2(\cos\tilde\theta)\tilde r^4 \sin\tilde\theta \, d\tilde r \, d\tilde\theta}{R_\ominus^2 \int_0^\pi \int_0^{R(\theta)} \tilde r^2 \sin\tilde\theta \, d\tilde r \, d\tilde\theta} \tag{3.33}$$

where $R(\tilde\theta) = R_\ominus - (R_\ominus - R_\odot)\cos^2\tilde\theta$. The $\tilde r$ integration reduces this to

$$-\frac{3}{5}\frac{\int_0^\pi P_2(\cos\tilde\theta) R_\ominus^5 \left(1 - 5\frac{R_\ominus - R_\odot}{R_\ominus}\cos^2\tilde\theta + \cdots\right)\sin\tilde\theta \, d\tilde\theta}{R_\ominus^2 \int_0^\pi R_\ominus^3 \left(1 - 3\frac{R_\ominus - R_\odot}{R_\ominus}\cos^2\tilde\theta + \cdots\right)\sin\tilde\theta \, d\tilde\theta}$$

$$\simeq \frac{3}{2}\frac{R_\ominus - R_\odot}{R_\ominus}\int_0^\pi P_2(\cos\tilde\theta)\cos^2\tilde\theta \sin\tilde\theta \, d\tilde\theta \simeq \frac{2}{5}\left(1 - \frac{R_\odot}{R_\ominus}\right) \tag{3.34}$$

since

$$\int_0^\pi P_2(\cos\tilde\theta)\cos^2\tilde\theta\sin\tilde\theta \, d\tilde\theta = \frac{4}{15} \tag{3.35}$$

□

Substituting (3.32) into (3.31) yields the following approximate formula for the amount of flattening

$$R_\ominus - R_\odot = \frac{5\Omega^2 R_\ominus^4}{4M} \simeq 27.2 \text{ km} \tag{3.36}$$

The discrepancy with the observed value of 21.4 km is due to the aforementioned uniform-density assumption.

3.3.0.2 Torque forces

When an oblate body (such as Earth) is placed in the gravitational field of a point-like perturber (Sun, in this case), it experiences (in addition to the regular gravitational pull) a torque perpendicular to its axis. This will not affect Earth's orbital motion, but it causes precession and nutation of the axis - a rather important phenomenon.

To find the magnitude of this torque, we expand Sun's gravitational (point-like) force, placing the origin of this expansion at the center of Earth. This yields the same result as (3.15) and (3.17), with $m \to 0$ (infinitesimal mass of a test particle). The corresponding torque is thus

$$\frac{3M_s}{R_s^5}(\mathbf{r}\cdot\mathbf{R}_s)(\mathbf{r}\times\mathbf{R}_s) \tag{3.37}$$

yet to be integrated over Earth's volume and multiplied by Earth's mass density (assuming this to be *uniform*, we get, to a good approximation,

$$\frac{m_e}{\frac{4}{3}\pi R_\ominus^3} \tag{3.38}$$

where m_e is Earth's *regular* mass).

In spherical coordinates (Earth's axis defining the z direction), $(\mathbf{r}\cdot\mathbf{R}_s)(\mathbf{r}\times\mathbf{R}_s)$ can be simplified (when further multiplied by 2π, to account for the trivial φ integration) to yield

$$-\frac{\pi}{2}r^2(1+3\cos 2\theta)\mathfrak{j}\mathbf{R}_z(\mathbf{R}_x+\mathfrak{i}\mathbf{R}_y) \tag{3.39}$$

Note that we have started *assuming* that two adjacent vectors —or quaternions in general— are multiplied in the usual quaternionic manner, without an explicit use of the '∘' symbol; this symbol is employed only occasionally, for extra emphasis, from now on.

Then, since

$$\int_0^{R_\ominus+(R_\odot-R_\ominus)\cos^2\theta} r^4\mathrm{d}r \simeq \frac{R_\ominus^5}{5}+R_\ominus^4(R_\odot-R_\ominus)\cos^2\theta \tag{3.40}$$

and

$$\int_0^\pi\left(\frac{R_\ominus^5}{5}+R_\ominus^4(R_\odot-R_\ominus)\cos^2\theta\right)(1+3\cos 2\theta)\sin\theta\,\mathrm{d}\theta=\frac{16}{15}R_\ominus^4(R_\odot-R_\ominus) \tag{3.41}$$

the resulting torque is

$$\frac{8M_s m_e}{5R_s^5}R_\ominus(R_\ominus-R_\odot)\mathfrak{j}\mathbf{R}_z(\mathbf{R}_x+\mathfrak{i}\mathbf{R}_y)=\frac{4M_s m_e}{R_s^5}J_2R_\ominus^2\mathfrak{j}\mathbf{R}_z(\mathbf{R}_x+\mathfrak{i}\mathbf{R}_y) \tag{3.42}$$

Taking

$$\mathbf{R}_s=R_s(\mathfrak{k}\cos\omega_s+\mathfrak{j}\cos\Theta\sin\omega_s+\mathfrak{i}\sin\Theta\sin\omega_s) \tag{3.43}$$

for Sun's 'orbit' (assumed circular), and averaging (3.42) over many such orbits (i.e. over ω_s) yields

$$\frac{2M_s m_e}{R_s^3}J_2R_\ominus^2\mathfrak{k}\sin\Theta\cos\Theta \tag{3.44}$$

where $\Theta\simeq 23^\circ 27'$ is Earth's obliquity (the angle between its axis and its orbit's normal). Note that the torque is perpendicular to the direction of Earth's axis, and oriented as if trying to 'straighten out' this axis (made it perpendicular to the ecliptic). This is opposite to the regular spinning top, where the torque caused by gravity is acting as if attempting to 'tip it over'. Instead of changing the axis' obliquity (true in both cases) precession around the perpendicular direction ensues (in the case of Earth, this precession is *retrograde*).

There will be a similar contribution to this torque from Moon's gravitational potential, to a good approximation given by the previous formula, with M_s and R_s replaced by M_m and R_m respectively (the gravitational mass and average distance of Moon - note that this makes Moon's effect more than double of that caused by Sun).

From the theory of spinning top it follows that the precession speed is given by the torque's magnitude, divided by the product of $\sin\Theta$, Earth's moment of inertia ($\simeq 2/5\ m_eR_\ominus^2$) and Earth's rotational speed Ω, i.e.

$$\frac{5}{\Omega}\left(\frac{M_s}{R_s^3}+\frac{M_m}{R_m^3}\right)J_2\cos\Theta\simeq 8.61\times 10^{-12}\ \mathrm{sec}^{-1} \tag{3.45}$$

This translates into one precession cycle every 23.1 millennia - reasonably close to the correct value of 25.4 millennia, considering the uniform-density assumption (and other approximations).

3.3.1 Tidal forces

In this segment we consider a 'liquid' (capable of quickly modifying its shape) primary with a relatively large satellite (for simplicity, we assume its orbit to be circular). Under these circumstances, the satellite is capable of distorting the primary's shape, stretching it in a football-like manner in the satellite's direction (both towards and away from the satellite, at the near and far side, respectively). The amount and shape of this stretching (let us denote Δ the corresponding *radial* increase) is given by the following function of θ (the angle between the direction where the stretching takes place, and that of the satellite)

$$\Delta = \frac{5}{8}\frac{mR^4}{Mr^3}(3\cos 2\theta - 1) \tag{3.46}$$

Proof. We will assume that the primary is non-rotating and therefore perfectly spherical. For an oblate primary (the usual case), the effect we derive below is simply superimposed on the existing shape.

Let us introduce a (moving) coordinate system with the origin at the primary's center, the z axis pointing to the satellite, and the x axis in its orbital plane ($R+\Delta$, θ and φ are spherical coordinates of the primary's surface point, R being its original radius). The overall gravitational potential at the surface is given by

$$V_{\text{gr}} = -\frac{M}{R+\Delta} + \frac{M\ J_2R^2(3\cos^2\theta - 1)}{2(R+\Delta)^3} - \frac{m}{\sqrt{r^2 + (R+\Delta)^2 + 2r(R+\Delta)\cos\theta}} \tag{3.47}$$

where r is the distance between the two bodies, and J_2 (satellite induced, its value yet to be determined) is measured with respect to the new coordinates.

From the gravitational potential, we need to subtract the potential of rotational motion of the two bodies around their center of mass, located at $rm/(M+m)$ from the primary's center. Since the bodies themselves do not rotate, yet another term is needed to 'un-rotate' the primary, thus:

$$\begin{aligned} V_{\text{rot}} = &-\frac{M+m}{2r^3}\left[(R+\Delta)^2\sin^2\theta\cos^2\varphi + \left((R+\Delta)\cos\theta + \frac{mr}{M+m}\right)^2\right] \\ &+\frac{M+m}{2r^3}\left[(R+\Delta)^2\sin^2\theta\cos^2\varphi + (R+\Delta)^2\cos^2\theta\right] \end{aligned} \tag{3.48}$$

which is based on (1.61), with $\Omega = \sqrt{(M+m)/r^3}$. Note that the two sets of square brackets contain the second power of the distance of a surface particle form the respective axis of rotation (the first running through the center of mass, the second through the primary's center).

Expanding the total potential in terms of Δ/R, R/r and J_2, we get

$$k\frac{M}{R^2} + \frac{3}{2}\left(\frac{M\ J_2}{R} - \frac{mR^2}{r^3}\right)\cos^2\theta + \frac{M}{R^2}\Delta \tag{3.49}$$

where kM/R^2 is a constant which stands for all θ-independent terms (the factor M/R^2 has been included only for future convenience).

This implies that

$$\Delta = \frac{3}{2}\left(\frac{mR^4}{Mr^3} - RJ_2\right)(\cos^2\theta + k) \tag{3.50}$$

Computing J_2 with the help of (3.33) and (3.35), and using the uniform-density assumption, yields

$$J_2 = -\frac{3}{2}\int_0^\pi P_2(\cos\theta)\frac{\Delta}{R}\sin\theta\ \mathrm{d}\theta = -\frac{3}{5}\left(\frac{mR^3}{Mr^3} - J_2\right) \tag{3.51}$$

Solving for J_2, we obtain

$$J_2 = -\frac{3}{2}\frac{m}{M}\left(\frac{R}{r}\right)^3 \tag{3.52}$$

which, when substituted back into (3.50) implies

$$\Delta = \frac{15}{4}\frac{mR^4}{Mr^3}(\cos^2\theta + k) \tag{3.53}$$

Not to change the primary's volume, i.e. to have $\int_0^\pi \Delta \sin\theta \, d\theta = 0$, we have to take $k = -1/3$. □

As the primary is usually rotating underneath its raised 'oceans', the corresponding friction causes the direction of the tidal stretching to be slightly displaced from the satellite's direction. Making the simplifying assumption that the primary's axis of rotation is perpendicular (or nearly so) to the satellite's orbiting plane, the angle of this displacement (say δ) will be proportional to the primary's rotational speed, relative to the satellite's angular velocity. Note that δ can be of either sign.

The primary, being distorted in this manner becomes, according to (3.26), a source of the following perturbation of the satellite's motion

$$\varepsilon\mathbf{f} = \frac{3mR_\ominus^5}{2r^8}\left(\frac{3}{2}\mathbf{r} - \frac{15}{2}\frac{(\mathbf{r}\cdot\mathbf{u})^2\mathbf{r}}{r^2} + 3(\mathbf{r}\cdot\mathbf{u})\mathbf{u}\right) \simeq -\frac{9mR_\ominus^5}{2r^8}(\mathbf{r} - \delta\mathbf{r}\mathrm{i}) \tag{3.54}$$

since $\mathbf{u}$ now equals to $\widehat{\mathbf{r}} \equiv \mathbf{r}/r$ rotated by the angle δ (assumed small), i.e. $\mathbf{u} \simeq \widehat{\mathbf{r}}(1 + \delta\mathrm{i})$ and $\mathbf{r}\cdot\mathbf{u} \simeq r$.

In the last chapter we show that the δ-term causes a slow secular increase (decrease) of the semimajor axis when the rotational speed is bigger (smaller) than the satellite's orbital speed.

3.4 MISCELLANEOUS

In this section, we briefly mention most of the remaining possible perturbing forces [2], [13].

3.4.1 Drag

This effect is due to the satellite passing through Earth's atmosphere, even though usually only in relatively high altitudes. The force has a direction opposite to the satellite's velocity, and a magnitude which is to a good approximation proportional to the satellite's speed squared, its cross-section area C, and the local air density ρ (we take this to be a function of r only) i.e.

$$\varepsilon\mathbf{f} = -\varepsilon\frac{\mu}{m}C\frac{\mathbf{r}'\,|\mathbf{r}'|}{4ar^2}\rho(r) \tag{3.55}$$

where ε is a dimensionless constant, and m is now the *regular* (not gravitational) mass of the satellite. The division by $4ar^2/\mu$ facilitates the $\dot{\mathbf{r}} \to \mathbf{r}'$ conversion.

In the first iteration, the *Kepler-frame* version of this force reads

$$-\varepsilon\frac{\mu}{m}C\frac{\mathrm{j}\left(z - \frac{\beta^2}{z}\right)\sqrt{1+\beta^4-\beta^2(z^2+\frac{1}{z^2})}}{a\left(1+\beta^2+\beta(z+\frac{1}{z})\right)^2}\rho\left(a + \frac{a\beta}{1+\beta^2}(z+\frac{1}{z})\right)$$

$$\simeq -\varepsilon\frac{\mu}{ma}C\mathrm{j}\left[z\rho(a) - (1+z^2)\left(2\rho(a) - a\rho'(a)\right)\beta + \cdots\right] \tag{3.56}$$

when expanded in β.

3.4.2 Kepler shear

When a large number of small bodies (called, in this context, 'particles') orbits the same primary, they will occasionally collide with one another. When a collision occurs at a particle's apocenter, the particle (moving unusually slowly compared to other particles nearby) will be more likely hit from behind, its speed increasing; similarly, at its pericenter, the particle will be slowed down by most collisions. The net effect of many such collisions can be translated into the following perturbing force

$$\varepsilon\mathbf{f} = \frac{\mu\rho C}{m}\left(\frac{\mathbf{r}\circ\mathrm{i}}{r^{3/2}} - \frac{\mathbf{r}'}{2r\sqrt{a}}\right)\left|\frac{\mathbf{r}\circ\mathrm{i}}{r^{3/2}} - \frac{\mathbf{r}'}{2r\sqrt{a}}\right| \tag{3.57}$$

whose direction is given by the difference between the velocity of particles in a perfectly circular orbit at the satellite's location (also sharing its orbital plane) and the satellite's true velocity, and whose magnitude is the square of this difference. Similarly to drag, C and ρ are the satellite's cross-section area and the particles' 'density' (including the empty space in between), and m is the satellite's *regular* mass; the value of ρ can now be considered constant.

When evaluated with the unperturbed solution (good only for the first iteration), converted to Kepler's frame, and expanded in β, this force reads

$$\frac{\mu\rho C}{4am}\mathrm{j}\beta^2\sqrt{10 - 3z^2 - \frac{3}{z^2}}\left(3 - z^2 - \frac{3 - 2z^2 - 13z^4 + 8z^6}{1 - 3z^2}\beta + \cdots\right) \tag{3.58}$$

3.4.3 Radiation pressure

Both light and ionized gas (collectively called 'radiation') emanating from Sun (the primary) exerts pressure on an orbiting body. The resulting perturbing force is proportional to the relative velocity of the radiation with respect to the perturbed body, i.e.

$$v\frac{\mathbf{r}}{r} - \frac{\mathbf{r}'}{2r}\sqrt{\frac{\mu}{a}} \tag{3.59}$$

(where v is the radiation speed), further multiplied by the corresponding magnitude (to a good approximation equal to v itself), the radiation mass density κ/r^2 (κ is a constant with dimensions of kg m^{-1}; for light, 'mass' means energy divided by ν^2) and the body's cross-section area C. The resulting perturbing force is thus

$$\varepsilon\mathbf{f} = \frac{C\kappa v}{mr^2}\left(v\frac{\mathbf{r}}{r} - \frac{\mathbf{r}'}{2r}\sqrt{\frac{\mu}{a}}\right) \tag{3.60}$$

where m is again the *regular* mass of the perturbed body. This formula assumes that radiation is fully absorbed by the orbiting body. If it is at least partially reflected, one would need yet another (dimensionless) factor to make the corresponding adjustment to (3.60).

The first term (in parentheses) is easy to deal with, since it is proportional to the main gravitational force $\mu\mathbf{r}_{\mathrm{o}}/r^3$; thus it only modifies the effective value of μ (even though, for micrometer-sized objects, this force may become bigger than the gravitational pull itself, making the adjusted μ *negative*; these objects are then swept out of the solar system). The second term represents a braking force similar to drag (note that its magnitude decreases with the $\frac{5}{2}$ power of the distance from Sun), and causes the so-called POYNTING-ROBERTSON effect. Both of these forces are important only for orbiting bodies of a relatively small size (not more than 10 meters in diameter).

The Kepler-frame version of Poynting-Robertson force, expanded in β, reads

$$-\frac{C\kappa v}{2m}\sqrt{\frac{\mu}{a}}\frac{\mathbf{r}'_{\mathrm{o}}}{r^3} \simeq -\frac{C\kappa v}{ma^2}\sqrt{\frac{\mu}{a}}\mathrm{j}\left(z - 3\beta(1 + z^2) + \cdots\right) \tag{3.61}$$

revealing its drag-like character.

So far, we have assumed Sun itself to be the primary; that of course may not always be the case. When a satellite is orbiting a planet, (3.60) becomes (ignoring its second, in this context practically negligible, term)

$$\varepsilon \mathbf{f} = \frac{C\kappa v^2 \mathbf{R}}{mR^3} \tag{3.62}$$

where $\mathbf{R}$ (now nearly constant) is the location of the satellite with respect to Sun's center.

3.4.4 Yarkovsky effect

This is caused by Sun's light (meaning electromagnetic radiation of any wave-length) being absorbed and then out-radiated by a *rotating* perturbed body. It is due to the fact that the Sun-facing side of the body is warmer then the 'night' side, thus emitting more heat. This at first appears equivalent to the previously-discussed radiation effect (with *reflected* light), except now the 'reflection' does not happen instantaneously, but has a specific time delay. Due to the body's rotation, the emitted heat's average direction corresponds to Sun's position *rotated* by an angle δ (equal to the time delay, multiplied by angular speed of the body's rotation). The perturbing force is the corresponding 'recoil', namely

$$\varepsilon \mathbf{f} = \varepsilon \frac{C\kappa v^2}{mr^3} e^{-\delta \mathbf{w}/2} \circ \mathbf{r} \circ e^{\delta \mathbf{w}/2} \tag{3.63}$$

where $\mathbf{w}$ is the *unit* direction of the satellite's rotational axis, $\mathbf{r}$ is its location with respect to Sun's center (replace $\mathbf{r}$ by $\mathbf{R}$ when Sun is *not* the primary), and ε is a small dimensionless constant which depends on the body's shape, size and composition (e.g. when the body is so small that it heats up quickly and *uniformly*, ε will tend to zero), and which also reflects the fact that not all impinging light is absorbed and converted to heat.

Assuming that δ and β are small, and that Sun *is* the primary, we can expand the Kepler-frame version of (3.63) as follows:

$$\varepsilon \frac{C\kappa v^2}{ma^2} \left(\mathfrak{k} z - \mathfrak{k} \beta (1 + 3z^2) + \mathfrak{j} w_z \delta z + w_x \frac{\delta}{2} (z - \frac{1}{z}) - \mathfrak{i} w_y \frac{\delta}{2} (z + \frac{1}{z}) + \ldots \right) \tag{3.64}$$

where the first two terms in parentheses represent the regular radiation pressure, and the remaining terms yield the YARKOVSKY effect. When the rotational axis is perpendicular to the orbit's plane, only the w_z-proportional term remains (again, clearly of drag-like nature).

3.4.5 Relativistic corrections

These are most conspicuous in the case of a Sun-orbiting planet. There is a general formula for relativistic corrections proportional to $1/c^2$ (the EINSTEIN-INFELD-HOFFMANN equation) applicable to an arbitrary many-body system (c is the speed of light). This formula reduces, when neglecting the remaining planets, and ignoring terms proportional to Sun's velocity (small compared to the planet's velocity $\dot{\mathbf{r}}$), to

$$\frac{\mu}{c^2} \left(4\mu \frac{\mathbf{r}}{r^4} + 4\dot{\mathbf{r}} \frac{\dot{\mathbf{r}} \cdot \mathbf{r}}{r^3} - \mathbf{r} \frac{\dot{\mathbf{r}} \cdot \dot{\mathbf{r}}}{r^3} \right) = \frac{\mu^2}{c^2} \left(4 \frac{\mathbf{r}}{r^4} + \mathbf{r}' \frac{\mathbf{r}' \cdot \mathbf{r}}{ar^5} - \mathbf{r} \frac{\mathbf{r}' \cdot \mathbf{r}'}{4ar^5} \right) \tag{3.65}$$

The usual Kepler-frame, β-expanded form is

$$\frac{\mu^2}{c^2 a^3} \mathfrak{k} \left(3z - \beta(1 + 15z^2) + \cdots \right) \tag{3.66}$$

CHAPTER 4

Solar system

Abstract

We compute the effect of perturbations of the type discussed in Section 3.1. Since the masses of all planets are small (compared to Sun's mass), only *first-order* effects are considered.[1]

We also utilize yet another approximation, facilitated by the fact that the resulting changes (circulation, libration and nutation) of the orbital elements are substantially (several orders of magnitude) slower than the orbital periods of the perturbing planets. As a result, the time-dependent coefficients of the $\mathbb{Q}$ and $\mathbb{W}$ do not affect the perturbed planet's long-run behavior and can be dropped out of the corresponding differential equations. This is equivalent to replacing each perturbing force by its long-run *average* (computed by averaging over one period of each of the perturbing planets). The perturbing forces thus become effectively time-independent, implying that only the autonomous-case formulas will be required in this chapter.

In general, one must be cautious when applying this approximation, and refrain from using it when any one of the perturbing forces is COMMENSURABLE with the motion of the perturbed planet (meaning that the two periods are in a simple ratio, such as 2:5).

4.1 PERTURBING FORCES

According to (3.3), the perturbing force is now a sum of several such small forces, namely

$$\sum_{\ell} \varepsilon_\ell \mathbf{f}_\ell = -\mu \sum_{\ell} \varepsilon_\ell \left(\frac{\mathbf{r} - \mathbf{r}_\ell}{|\mathbf{r} - \mathbf{r}_\ell|^3} + \frac{\mathbf{r}_\ell}{r_\ell^3} \right) \tag{4.1}$$

where

$$\mathbf{r}_\ell = \overline{\mathbb{R}}_\ell \circ \mathfrak{k} \frac{a_\ell}{1+\beta_\ell^2} \Big(\exp(\mathrm{i}\omega_\ell) + 2\beta_\ell + \beta_\ell^2 \exp(-\mathrm{i}\omega_\ell) \Big) \circ \mathbb{R}_\ell \tag{4.2}$$

ω_ℓ being the eccentric anomaly, measured from the respective aphelion, in exact analogy to (1.59), and each $\mathbb{R}_\ell$ is parametrized by

$$\exp(\mathrm{i}\frac{\psi_\ell}{2}) \circ \exp(\mathfrak{k}\frac{\theta_\ell}{2}) \circ \exp(\mathrm{i}\frac{\phi_\ell}{2}) \tag{4.3}$$

Since first-order corrections (to the rate of change of orbital elements, and to the orbit's parallel and perpendicular distortions) are linear in ε_ℓ, the total correction is just a sum of the corresponding individual corrections; this implies that we need to consider only *one* perturbing planet at a time.

4.1.1 Kepler-frame conversion

Expressing $\mathbf{r}_\ell$ in Kepler's frame of the perturbed planet (we denote the answer $\mathbf{r}_{\ell o}$) requires replacing $\mathbb{R}_\ell$ in (4.2) by $\mathbb{R}_{\ell o} = \mathbb{R}_\ell \circ \overline{\mathbb{R}}$ ($\mathbb{R}$ being the attitude of $\mathbf{r}$). The *result* is then expressed in terms of Euler angles, thus

$$\mathbb{R}_{\ell o} \equiv \exp(\mathrm{i}\frac{\psi_{\ell o}}{2}) \circ \exp(\mathfrak{k}\frac{\theta_{\ell o}}{2}) \circ \exp(\mathrm{i}\frac{\phi_{\ell o}}{2}) \tag{4.4}$$

[1]Compare with [8], [28], [29] and [47].

Jan Vrbik

where

$$\theta_{\ell o} = \arccos\left[\cos\theta_\ell \cos\theta + \sin\theta_\ell \sin\theta \cos(\phi_\ell - \phi)\right] \tag{4.5a}$$

$$\phi_{\ell o} = -\psi + \arctan\left[\cos\theta \cos(\phi_\ell - \phi) - \sin\theta \cot\theta_\ell,\ \sin(\phi_\ell - \phi)\right] \tag{4.5b}$$

$$\psi_{\ell o} = \psi_\ell + \arctan\left[-\cos\theta_\ell \cos(\phi_\ell - \phi) + \sin\theta_\ell \cot\theta,\ -\sin(\phi_\ell - \phi)\right] \tag{4.5c}$$

Proof. To verify these, one has to solve:

$$\mathbb{R}_\ell \circ \overline{\mathbb{R}} \equiv \mathbb{T} = \exp(\mathfrak{i}\frac{\psi_{\ell o}}{2}) \circ \exp(\mathfrak{k}\frac{\theta_{\ell o}}{2}) \circ \exp(\mathfrak{i}\frac{\phi_{\ell o}}{2}) \tag{4.6}$$

for $\phi_{\ell o}$, $\theta_{\ell o}$ and $\psi_{\ell o}$. The last equation implies that

$$\begin{aligned} &\exp(-\mathfrak{i}\frac{\psi}{2}) \circ \mathbb{T}^{(\mathfrak{i})} \circ \mathbb{T} \circ \exp(\mathfrak{i}\frac{\psi}{2}) = \exp\left(-\mathfrak{i}\frac{\phi_{\ell o}+\psi}{2}\right) \circ \exp(\mathfrak{k}\,\theta_{\ell o}) \circ \exp\left(\mathfrak{i}\frac{\phi_{\ell o}+\psi}{2}\right) \\ &= \cos\theta_{\ell o} + \mathfrak{k}\sin\theta_{\ell o} \circ \left[\cos(\phi_{\ell o}+\psi) + \mathfrak{i}\,\sin(\phi_{\ell o}+\psi)\right] \end{aligned} \tag{4.7}$$

and

$$\begin{aligned} &\exp(-\mathfrak{i}\frac{\psi_\ell}{2}) \circ \mathbb{T} \circ \mathbb{T}^{(\mathfrak{i})} \circ \exp(\mathfrak{i}\frac{\psi_\ell}{2}) = \exp\left(\mathfrak{i}\frac{\psi_{\ell o}-\psi_\ell}{2}\right) \circ \exp(\mathfrak{k}\,\theta_{\ell o}) \circ \exp\left(-\mathfrak{i}\frac{\psi_{\ell o}-\psi_\ell}{2}\right) \\ &= \cos\theta_o + \mathfrak{k}\sin\theta_o \circ \left[\cos(\psi_{\ell o}-\psi_\ell) - \mathfrak{i}\,\sin(\psi_{\ell o}-\psi_\ell)\right] \end{aligned} \tag{4.8}$$

where $\mathbb{T}^{(\mathfrak{i})}$ denotes the $\mathfrak{i}$-conjugate of $\mathbb{T}$ (changing the sign of its $\mathfrak{i}$ component only). Note that, similarly to the full conjugate,

$$(\mathbb{A} \circ \mathbb{B})^{(\mathfrak{i})} = \mathbb{B}^{(\mathfrak{i})} \circ \mathbb{A}^{(\mathfrak{i})} \tag{4.9}$$

Correspondingly,

$$\begin{aligned} &\exp(-\mathfrak{i}\frac{\psi}{2}) \circ \overline{\mathbb{R}}^{(\mathfrak{i})} \circ \mathbb{R}_\ell^{(\mathfrak{i})} \circ \mathbb{R}_\ell \circ \overline{\mathbb{R}} \circ \exp(\mathfrak{i}\frac{\psi}{2}) \\ &= \exp(-\mathfrak{k}\frac{\theta}{2}) \circ \exp\left(-\mathfrak{i}\,\frac{\phi_\ell-\phi}{2}\right) \circ \exp(\mathfrak{k}\,\theta_\ell) \circ \exp\left(\mathfrak{i}\frac{\phi_\ell-\phi}{2}\right) \circ \exp(-\mathfrak{k}\frac{\theta}{2}) \\ &= \left[\cos\theta_\ell + \mathfrak{k}\sin\theta_\ell\cos(\phi_\ell-\phi)\right] \circ (\cos\theta - \mathfrak{k}\sin\theta) + \mathfrak{j}\sin\theta_\ell\sin(\phi_\ell-\phi) \\ &= \cos\theta_\ell\cos\theta + \sin\theta_\ell\sin\theta\cos(\phi_\ell-\phi) - \mathfrak{k}\left[\cos\theta_\ell\sin\theta - \sin\theta_\ell\cos\theta\cos(\phi_\ell-\phi)\right] + \mathfrak{k}\sin\theta_\ell\sin(\phi_\ell-\phi)\mathfrak{i} \end{aligned} \tag{4.10}$$

which implies (4.5a) and (4.5b), and

$$\begin{aligned} &\exp(-\mathfrak{i}\frac{\psi_\ell}{2}) \circ \mathbb{R}_\ell \circ \overline{\mathbb{R}} \circ \overline{\mathbb{R}}^{(\mathfrak{i})} \circ \mathbb{R}_\ell \circ \exp(\mathfrak{i}\frac{\psi_\ell}{2}) \\ &= \exp(\mathfrak{k}\frac{\theta_\ell}{2}) \circ \exp\left(\mathfrak{i}\frac{\phi_\ell-\phi}{2}\right) \circ \exp(-\mathfrak{k}\theta) \circ \exp\left(-\mathfrak{i}\,\frac{\phi_\ell-\phi}{2}\right) \circ \exp(\mathfrak{k}\frac{\theta_\ell}{2} \\ &= \left[\cos\theta - \mathfrak{k}\sin\theta\cos(\phi_\ell-\phi)\right] \circ (\cos\theta_\ell + \mathfrak{k}\sin\theta_\ell) + \mathfrak{j}\sin\theta\sin(\phi_\ell-\phi) \\ &= \cos\theta_\ell\cos\theta + \sin\theta_\ell\sin\theta\cos(\phi_\ell-\phi) + \mathfrak{k}\left[\cos\theta\sin\theta_\ell - \sin\theta\cos\theta_\ell\cos(\phi_\ell-\phi)\right] + \mathfrak{k}\sin\theta\sin(\phi_\ell-\phi)\mathfrak{i} \end{aligned} \tag{4.11}$$

which confirms (4.5a) and implies (4.5c). □

4.1.2 Long-run averaging

We will now replace the perturbing force by its single-period average. In general, averaging of any quantity $g(\omega_\ell)$ is facilitated by

$$\frac{1}{2\pi}\int_0^{2\pi} g(\omega_\ell)\frac{r_\ell}{a_\ell}\,\mathrm{d}\omega_\ell \equiv \overline{g} \tag{4.12}$$

according to (1.28b) and (1.61), where

$$r_\ell = a_\ell(1+\frac{2\beta_\ell}{1+\beta_\ell}\cos\omega_\ell) \tag{4.13}$$

Note that this represents averaging in *regular* time, even though, more properly, we should be using the *modified* time of the perturbed planet as independent variable. Luckily, the two averages are identical (in a long run), assuming that the corresponding frequencies are not commensurable.

4.1.3 Expanding in β, β_ℓ, and $\theta_{\ell\mathrm{o}}$

In addition to such averaging, one can further substantially simplify both the algebra and the results of these computations by expanding all quantities in terms of β, β_ℓ, and $\theta_{k\mathrm{o}}$ (keeping their linear contributions only). This assumes that all planets have small eccentricities, and that their orbital planes are not much inclined relative to the ecliptic (the plane of Earth's orbit).

To carry out the such an expansion of the perturbing force, we first note that, to this approximation,

$$\mathbb{R}_{\ell\mathrm{o}} \simeq \exp\left(\mathrm{i}\frac{\phi_{\ell\mathrm{o}}+\psi_{\ell\mathrm{o}}}{2}\right)+\mathfrak{k}\frac{\theta_{\ell\mathrm{o}}}{2}\circ\exp\left(\mathrm{i}\frac{\phi_{\ell\mathrm{o}}-\psi_{\ell\mathrm{o}}}{2}\right)+\cdots \tag{4.14}$$

$$\mathbf{r}_{\ell\mathrm{o}} \simeq \overline{\mathbb{R}}_{\ell\mathrm{o}}\mathfrak{k}\,a_\ell\,(z_\ell+2\beta_\ell)\,\mathbb{R}_{\ell\mathrm{o}} = \mathfrak{k}\,a_\ell\,(z_\ell+2\beta_\ell)\circ\exp[\mathrm{i}\,(\phi_{\ell\mathrm{o}}+\psi_{\ell\mathrm{o}})]+\mathrm{i}\,\theta_{\ell\mathrm{o}}a_\ell\sin(\omega_\ell+\psi_{\ell\mathrm{o}})+\cdots \tag{4.15}$$

where $z_\ell \equiv \exp(\mathrm{i}\omega_\ell)$,

$$\mathbf{r}_\mathrm{o} \simeq \mathfrak{k}a(z+2\beta)+\cdots \tag{4.16}$$

$$\mathbf{r}_\mathrm{o}\circ\mathbf{r}_{\ell\mathrm{o}} \simeq -aa_\ell\left(\frac{z_\ell+2\beta_\ell}{z}+2\beta z_\ell\right)\exp[\mathrm{i}\,(\phi_{\ell\mathrm{o}}+\psi_{\ell\mathrm{o}})]+\mathrm{j}\,\theta_{\ell\mathrm{o}}aa_\ell z\sin(\omega_\ell+\psi_{\ell\mathrm{o}})+\cdots \tag{4.17}$$

$$\begin{aligned}|\,\mathbf{r}_\mathrm{o}-\mathbf{r}_{\ell\mathrm{o}}\,|^2 &\simeq r^2+r_\ell^2+\mathbf{r}_\mathrm{o}\mathbf{r}_{\ell\mathrm{o}}+\mathbf{r}_{\ell\mathrm{o}}\mathbf{r}_\mathrm{o} = a^2(1+4\beta\cos\omega)+a_\ell^2(1+4\beta_\ell\cos\omega_\ell)\\ &-2aa_\ell\cos(\omega_\ell-\omega+\phi_{\ell\mathrm{o}}+\psi_{\ell\mathrm{o}})-4aa_\ell\beta\cos(\omega_\ell+\phi_{\ell\mathrm{o}}+\psi_{\ell\mathrm{o}})-4aa_\ell\beta_\ell\cos(\omega-\phi_{\ell\mathrm{o}}-\psi_{\ell\mathrm{o}})\end{aligned} \tag{4.18}$$

which further implies that

$$\begin{aligned}\frac{1}{|\,\mathbf{r}_\mathrm{o}-\mathbf{r}_{\ell\mathrm{o}}\,|^3} &\simeq \frac{1}{\left(a^2+a_\ell^2-2aa_\ell\cos(\omega_\ell-\omega+\phi_{\ell\mathrm{o}}+\psi_{\ell\mathrm{o}})\right)^{3/2}}\\ &-6\frac{\beta a^2\cos\omega+a_\ell^2\beta_\ell\cos\omega_\ell-aa_\ell\beta\cos(\omega_\ell+\phi_{\ell\mathrm{o}}+\psi_{\ell\mathrm{o}})-aa_\ell\beta_\ell\cos(\omega-\phi_{\ell\mathrm{o}}-\psi_{\ell\mathrm{o}})}{\left(a^2+a_\ell^2-2aa_\ell\cos(\omega_\ell-\omega+\phi_{\ell\mathrm{o}}+\psi_{\ell\mathrm{o}})\right)^{5/2}}\\ &\equiv \frac{1}{\Delta^{3/2}}-6\frac{\beta a^2\cos\omega+a_\ell^2\beta_\ell\cos(\xi+\omega-\phi_{\ell\mathrm{o}}-\psi_{\ell\mathrm{o}})-aa_\ell\beta\cos(\xi+\omega)-aa_\ell\beta_\ell\cos(\omega-\phi_{\ell\mathrm{o}}-\psi_{\ell\mathrm{o}})}{\Delta^{5/2}}\end{aligned} \tag{4.19}$$

where $\xi \equiv \omega_\ell - \omega + \phi_{\ell o} + \psi_{\ell o}$ and $\Delta \equiv a^2 + a_\ell^2 - 2aa_\ell \cos\xi$.

Note that for the purpose of time averaging (see Formula 4.12),

$$\frac{1}{2\pi}\int_0^{2\pi} \cdots \mathrm{d}\omega_\ell \equiv \frac{1}{2\pi}\int_0^{2\pi} \cdots \mathrm{d}\xi \tag{4.20}$$

4.1.4 Integration formula

The the individual terms of the perturbing force are than averaged by utilizing

$$\frac{1}{2\pi}\int_0^{2\pi} \frac{\exp[\mathrm{i}\,(n\xi+\delta)]\,\mathrm{d}\xi}{\Delta^\alpha} = \frac{a_\mathrm{s}^{|n|}}{a_\mathrm{b}^{|n|+2\alpha}}\frac{\Gamma(\alpha+|n|)}{|n|!\Gamma(\alpha)}\,{}_2F_1\left(\alpha, |n|+\alpha, |n|+1; \frac{a_\mathrm{s}^2}{a_\mathrm{b}^2}\right)\exp(\mathrm{i}\delta) \equiv \mathsf{F}(\alpha,n)\exp(\mathrm{i}\delta) \tag{4.21}$$

where $a_\mathrm{s} = \min(a, a_\ell)$ and $a_\mathrm{b} = \max(a, a_\ell)$, n is an integer, and ${}_2F_1(\cdots)$ is the usual hypergeometric function. Note that the formula remains correct when we replace each exp(i ..) by either cos(..) or sin(..), and that, in our case, α is always a half-integer. One should also remember that each $\mathsf{F}(\alpha, n)$ is, implicitly, a function of a and a_ℓ.

Note that we have expressed the result in a form which makes it valid in the case of both $a_\ell < a$ (INNER perturbing planet) and $a_\ell > a$ (OUTER perturbing planet).

Furthermore, since

$$(\alpha-1)(a^2-a_\ell^2)^2\mathsf{F}(\alpha,n) = 2aa_\ell(\alpha-n-2)\mathsf{F}(\alpha-1,n+1) + (a^2+a_\ell^2)(\alpha+n-1)\mathsf{F}(\alpha-1,n)$$

we can always reduce (or increase) α by one unit (thus making it equal to 3/2, in more than one step if necessary; in this context, 3/2 is the most convenient value). Similarly, since

$$n(a^2+a_\ell^2)\mathsf{F}(\alpha,n) = aa_\ell\left[(\alpha+n-1)\mathsf{F}(\alpha,n-1) - (\alpha-n-1)\mathsf{F}(\alpha,n+1)\right] \tag{4.22}$$

we can always reduce or increase the value of n, until we reach a linear combination of $\mathsf{F}(\alpha, 1)$ and $\mathsf{F}(\alpha, 2)$. This implies that all integrals of type (4.21) can be expressed in terms of $\mathsf{F}(\frac{3}{2}, 1)$ and $\mathsf{F}(\frac{3}{2}, 2)$, coincidentally simplifying (often quite substantially) the complexity of resulting expressions.

Proof. Based on

$$\frac{\cos(nx)}{\Delta^{\alpha-1}} = \frac{(a^2+a_\ell^2-2aa_\ell\cos x)\cos(nx)}{\Delta^\alpha} = \frac{(a^2+a_\ell^2)\cos(nx) - aa_\ell\cos(nx+x) - aa_\ell\cos(nx-x)}{\Delta^\alpha} \tag{4.23}$$

we get

$$\mathsf{F}(\alpha-1,n) = (a^2+a_\ell^2)\mathsf{F}(\alpha,n) - aa_\ell\left[\mathsf{F}(\alpha,n+1)+\mathsf{F}(\alpha,n-1)\right] \tag{4.24}$$

Similarly

$$\int_0^{2\pi}\frac{n\cos(nx)dx}{\Delta^{\alpha-1}} = -\int_0^{2\pi}\frac{[\sin(nx)]'\,dx}{\Delta^{\alpha-1}} = \int_0^{2\pi}\frac{2aa_\ell(\alpha-1)\sin x\sin(nx)dx}{\Delta^\alpha}$$
$$= \int_0^{2\pi}\frac{aa_\ell(\alpha-1)[\cos(nx-x)-\cos(nx+x)]dx}{\Delta^\alpha} \tag{4.25}$$

implies

$$n\mathsf{F}(\alpha-1,n) = aa_\ell(\alpha-1)\left[\mathsf{F}(\alpha,n-1)-\mathsf{F}(\alpha,n+1)\right] \tag{4.26}$$

Substituting (4.24) into (4.26) eliminates $\mathsf{F}(\alpha-1,n)$, leading to (4.22).

Subtracting $(\alpha-1)$ times (4.24) from (4.26) yields

$$(n-\alpha+1)\mathsf{F}(\alpha-1,n) = -(\alpha-1)(a^2+a_\ell^2)\mathsf{F}(\alpha,n)+2aa_\ell(\alpha-1)\mathsf{F}(\alpha,n-1) \tag{4.27}$$

which implies

$$(n-\alpha+2)\mathsf{F}(\alpha-1,n+1) = -(\alpha-1)(a^2+a_\ell^2)\mathsf{F}(\alpha,n+1)+2aa_\ell(\alpha-1)\mathsf{F}(\alpha,n) \tag{4.28}$$

Multiplying (4.27) by $(a^2+a_\ell^2)$, and adding (4.28) multiplied by $2aa_\ell$, yields

$$(a^2+a_\ell^2)(n-\alpha+1)\mathsf{F}(\alpha-1,n)+2aa_\ell(n-\alpha+2)\mathsf{F}(\alpha-1,n+1) \tag{4.29}$$

on the left hand side side, and

$$\begin{aligned}&-(\alpha-1)(a^2-a_\ell^2)^2\mathsf{F}(\alpha,n)+2aa_\ell(\alpha-1)(a^2+a_\ell^2)\mathsf{F}(\alpha,n-1)-2aa_\ell(\alpha-1)(a^2+a_\ell^2)\mathsf{F}(\alpha,n+1)\\&=-(\alpha-1)(a^2-a_\ell^2)^2\mathsf{F}(\alpha,n)+2(a^2+a_\ell^2)n\mathsf{F}(\alpha-1,n)\end{aligned} \tag{4.30}$$

on the right hand side side; this proves (4.22). □

4.2 RESULTING EQUATIONS

We can now easily compute the required long-run averages:

$$\begin{aligned}\overline{\frac{1}{|\mathbf{r}_\text{o}-\mathbf{r}_{\ell\text{o}}|^3}} &\simeq \mathsf{F}(\tfrac{3}{2},0)+2\beta_\ell\mathsf{F}(\tfrac{3}{2},1)\cos(\omega-\phi_{\ell\text{o}}-\psi_{\ell\text{o}})\\&\quad-6\left(a^2\beta\cos\omega-aa_\ell\beta_\ell\cos(\omega-\phi_{\ell\text{o}}-\psi_{\ell\text{o}})\right)\left(\mathsf{F}(\tfrac{5}{2},0)-\frac{a_\ell}{a}\mathsf{F}(\tfrac{5}{2},1)\right)+\cdots\end{aligned} \tag{4.31}$$

$$\overline{\frac{\mathbf{r}_{\ell\text{o}}}{|\mathbf{r}_\text{o}-\mathbf{r}_{\ell\text{o}}|^3}} \simeq \mathrm{i}\theta_{\ell\text{o}}a_\ell\mathsf{F}(\tfrac{3}{2},1)\sin(\omega-\phi_{\ell\text{o}})+\cdots \tag{4.32}$$

$$\begin{aligned}\mathrm{Cx}\left(\overline{\frac{\mathbf{r}_\text{o}\circ\mathbf{r}_{\ell\text{o}}}{|\mathbf{r}_\text{o}-\mathbf{r}_{\ell\text{o}}|^3}}\right) &\simeq -aa_\ell\mathsf{F}(\tfrac{3}{2},1)-3a^2a_\ell\beta\left[\mathsf{F}(\tfrac{5}{2},0)\exp(-\mathrm{i}\omega)+\mathsf{F}(\tfrac{5}{2},2)\exp(\mathrm{i}\omega)\right]+6a^3a_\ell\beta\mathsf{F}(\tfrac{5}{2},1)\cos\omega\\&\quad-2aa_\ell\beta\mathsf{F}(\tfrac{3}{2},1)\exp(\mathrm{i}\omega)-6a^2a_\ell\beta_\ell\mathsf{F}(\tfrac{5}{2},1)\cos(\omega-\phi_{\ell\text{o}}-\psi_{\ell\text{o}})\\&\quad+3aa_\ell\beta_\ell\Big(\mathsf{F}(\tfrac{5}{2},0)\exp[\mathrm{i}(\phi_{\ell\text{o}}+\psi_{\ell\text{o}}-\omega)]+\mathsf{F}(\tfrac{5}{2},2)\exp[\mathrm{i}(\omega-\phi_{\ell\text{o}}-\psi_{\ell\text{o}})]\Big)\\&\quad-aa_\ell\beta_\ell\Big(3\mathsf{F}(\tfrac{3}{2},0)\exp[\mathrm{i}(\phi_{\ell\text{o}}+\psi_{\ell\text{o}}-\omega)]+\mathsf{F}(\tfrac{3}{2},2)\exp[\mathrm{i}(\omega-\phi_{\ell\text{o}}-\psi_{\ell\text{o}})]\Big)+\cdots\end{aligned} \tag{4.33}$$

and

$$\overline{\frac{\mathbf{r}_{\ell\text{o}}}{r_{\ell\text{o}}^3}} \simeq 0+\cdots \tag{4.34}$$

These enable us to find

$$\begin{aligned}\mathcal{Q}(z) &= -2\varepsilon_\ell\frac{a}{\mu(1+\beta z)}\mathrm{Cx}(\overline{\mathbf{r}_\text{o}\circ\mathbf{f}_\text{o}}) \simeq -2\varepsilon_\ell a(1-\beta z)\mathrm{Cx}\left(\overline{\frac{r^2+\mathbf{r}_\text{o}\circ\mathbf{r}_{\ell\text{o}}}{|\mathbf{r}_\text{o}-\mathbf{r}_{\ell\text{o}}|^3}}\right)\\&= -2\varepsilon_\ell a^3\overline{\frac{1}{|\mathbf{r}_\text{o}-\mathbf{r}_{\ell\text{o}}|^3}}-4\varepsilon_\ell a^3\beta\mathsf{F}(\tfrac{3}{2},0)(z+\frac{1}{z})+2\varepsilon_\ell a^3\beta\mathsf{F}(\tfrac{3}{2},0)z\\&\quad-2\varepsilon_\ell a\mathrm{Cx}\left(\overline{\frac{\mathbf{r}_\text{o}\circ\mathbf{r}_{\ell\text{o}}}{|\mathbf{r}_\text{o}-\mathbf{r}_{\ell\text{o}}|^3}}\right)-2\varepsilon a^2a_\ell\beta\mathsf{F}(\tfrac{3}{2},1)z+\cdots\end{aligned} \tag{4.35}$$

and

$$\mathcal{W}(z) = -4\varepsilon_\ell \frac{a}{\mu} r_\mathrm{o}\,\mathrm{Cx}\,(\overline{\mathbf{f}_\mathrm{o}}) \simeq -4\mathrm{i}\varepsilon_\ell a^2 a_\ell \theta_{\ell\mathrm{o}} \mathsf{F}(\tfrac{3}{2},1)\sin(\omega-\phi_{\ell\mathrm{o}}) + \cdots \tag{4.36}$$

from which we can easily establish the coefficients of individual powers of $z \equiv \exp(\mathrm{i}\omega)$, namely:

$$\frac{\mathcal{Q}_{-1}}{-2\varepsilon_\ell a} \simeq 2a^2\beta\mathsf{F}(\tfrac{3}{2},0) - 3a^2\beta\left((a^2+a_\ell)\mathsf{F}(\tfrac{5}{2},0) - 2aa_\ell\mathsf{F}(\tfrac{5}{2},1)\right)$$
$$+3aa_\ell\beta_\ell\left((a^2+a_\ell)\mathsf{F}(\tfrac{5}{2},0) - 2aa_\ell\mathsf{F}(\tfrac{5}{2},1)\right)\exp[\mathrm{i}(\phi_{\ell\mathrm{o}}+\psi_{\ell\mathrm{o}})]$$
$$-a^2\beta_\ell\left(3\frac{a_\ell}{a}\mathsf{F}(\tfrac{3}{2},0) - \mathsf{F}(\tfrac{3}{2},1)\right)\exp[\mathrm{i}(\phi_{\ell\mathrm{o}}+\psi_{\ell\mathrm{o}})] + \cdots \tag{4.37}$$

$$\frac{\mathcal{Q}_0}{-2\varepsilon_\ell a} \simeq a^2\mathsf{F}(\tfrac{3}{2},0) - aa_\ell\mathsf{F}(\tfrac{3}{2},1) + \cdots \tag{4.38}$$

$$\frac{\mathcal{Q}_1}{-2\varepsilon_\ell a} \simeq -3a^2\beta\left(a^2\mathsf{F}(\tfrac{5}{2},0) + a_\ell\mathsf{F}(\tfrac{5}{2},2) - 2aa_\ell\mathsf{F}(\tfrac{5}{2},1)\right) + a^2\beta\left(\mathsf{F}(\tfrac{3}{2},0) - \frac{a_\ell}{a}\mathsf{F}(\tfrac{3}{2},1)\right)$$
$$+a^2\beta_\ell\left(\mathsf{F}(\tfrac{3}{2},1) - \frac{a_\ell}{a}\mathsf{F}(\tfrac{3}{2},2)\right)\exp[-\mathrm{i}(\phi_{\ell\mathrm{o}}+\psi_{\ell\mathrm{o}})] + 3aa_\ell\beta_\ell\left(a^2\mathsf{F}(\tfrac{5}{2},0) + a_\ell\mathsf{F}(\tfrac{5}{2},2)\right.$$
$$\left.-2aa_\ell\mathsf{F}(\tfrac{5}{2},1)\right)\exp[-\mathrm{i}(\phi_{\ell\mathrm{o}}+\psi_{\ell\mathrm{o}})] + \cdots \tag{4.39}$$

$$\frac{\mathcal{W}_1}{-2\varepsilon_\ell a} \simeq aa_\ell\theta_{\ell\mathrm{o}}\mathsf{F}(\tfrac{3}{2},1)\exp(-\mathrm{i}\phi_{\ell\mathrm{o}}) + \cdots \tag{4.40}$$

$$\mathcal{W}_{-1} = \mathcal{W}_1^* \tag{4.41}$$

and the rest of them are, to our approximation, equal to zero.

4.2.1 Orbital-element derivatives

Utilizing these results, we now compute

$$a' \simeq 2a\mathrm{Im}(\mathcal{Q}_0 - \beta\mathcal{Q}_{-1}) = 0 + \cdots \tag{4.42a}$$

$$\beta' \simeq -\frac{1}{4}\mathrm{Im}(\mathcal{Q}_1 + 3\beta\mathcal{Q}_0 + 3\mathcal{Q}_{-1} + \beta\mathcal{Q}_{-2}) = \varepsilon_\ell a^2 a_\ell\beta_\ell\mathsf{F}(\tfrac{3}{2},2)\sin(\phi_{\ell\mathrm{o}}+\psi_{\ell\mathrm{o}}) + \cdots \tag{4.42b}$$

$$Z_1 \simeq -\frac{1}{2}\mathrm{Im}(\mathcal{W}_1 + \beta\mathcal{W}_0) = -\varepsilon_\ell a^2 a_\ell\theta_{\ell\mathrm{o}}\mathsf{F}(\tfrac{3}{2},1)\sin\phi_{\ell\mathrm{o}} + \cdots \tag{4.42c}$$

$$Z_2 \simeq -\frac{1}{2}\mathrm{Re}(\mathcal{W}_1) = \varepsilon_\ell a^2 a_\ell\theta_{\ell\mathrm{o}}\mathsf{F}(\tfrac{3}{2},1)\cos\phi_{\ell\mathrm{o}} + \cdots \tag{4.42d}$$

$$Z_3 \simeq \frac{1}{4\beta}\mathrm{Re}(-\mathcal{Q}_1 + \beta\mathcal{Q}_0 + 3\mathcal{Q}_{-1} + \beta\mathcal{Q}_{-2}) = \varepsilon_\ell a^2 a_\ell\left(\mathsf{F}(\tfrac{3}{2},1) - \frac{\beta_\ell}{\beta}\mathsf{F}(\tfrac{3}{2},2)\cos(\phi_{\ell\mathrm{o}}+\psi_{\ell\mathrm{o}})\right) + \cdots \tag{4.42e}$$

(s'_p is not needed, as we switch to regular time t to integrate these equations), where summation over ℓ is now implied on the right hand side, to combine the effect of all perturbing planets.

It is interesting to note that the β' and Z_3 expansions were originally quite lengthy, but after expressing all the $\mathsf{F}(\alpha,n)$'s in terms of $\mathsf{F}(\frac{3}{2},1)$ and $\mathsf{F}(\frac{3}{2},2)$, using

$$\mathsf{F}(\tfrac{3}{2},0) = \frac{2(a^2+a_\ell^2)}{3aa_\ell}\mathsf{F}(\tfrac{3}{2},1) - \frac{1}{3}\mathsf{F}(\tfrac{3}{2},2) \tag{4.43a}$$

$$3(a^2-a_\ell^2)^2\mathsf{F}(\tfrac{5}{2},0) = \frac{2(a^4+3a^2a_\ell^2+a_\ell^4)}{aa_\ell}\mathsf{F}(\tfrac{3}{2},1) - (a^2+a_\ell^2)\mathsf{F}(\tfrac{3}{2},2) \tag{4.43b}$$

$$3(a^2-a_\ell^2)^2\mathsf{F}(\tfrac{5}{2},1) = 10aa_\ell\mathsf{F}(\tfrac{3}{2},1) - (a^2+a_\ell^2)\mathsf{F}(\tfrac{3}{2},2) \tag{4.43c}$$

$$3(a^2-a_\ell^2)^2\mathsf{F}(\tfrac{5}{2},2) = 5(a^2+a_\ell^2)\mathsf{F}(\tfrac{3}{2},1) - 2aa_\ell\mathsf{F}(\tfrac{3}{2},2) \tag{4.43d}$$

an almost magical cancellation took place (as pointed out earlier).

4.2.1.1 Regular-time conversion

With the help of (1.19), we can finally present the corresponding formulas for the rate of change of the orbital elements in *regular time,* by multiplying each right hand sides by $\frac{1}{2}\sqrt{\mu/a^3} \simeq \pi/T$, T being the perturbed planet's orbital period (this follows from Eq. 1.61, discarding its β-proportional component, in accordance with our approximations). We get

$$\dot{a} = 0 \tag{4.44a}$$

$$\dot{\beta} = \varepsilon_\ell \tfrac{\pi}{T} a^2 a_\ell \mathsf{F}(\tfrac{3}{2}, 2) \sin(\phi_{\ell o} + \psi_{\ell o}) \beta_\ell \tag{4.44b}$$

$$\dot{\phi} = \varepsilon_\ell \tfrac{\pi}{T} a^2 a_\ell \mathsf{F}(\tfrac{3}{2}, 1) \cos(\phi_{\ell o} + \psi) \frac{\theta_{\ell o}}{\sin\theta} \tag{4.44c}$$

$$\dot{\psi} = \varepsilon_\ell \tfrac{\pi}{T} a^2 a_\ell \mathsf{F}(\tfrac{3}{2}, 1) - \varepsilon_\ell \tfrac{\pi}{T} a^2 a_\ell \mathsf{F}(\tfrac{3}{2}, 2) \cos(\phi_{\ell o} + \psi_{\ell o}) \frac{\beta_\ell}{\beta} - \dot{\phi}\cos\theta \tag{4.44d}$$

$$\dot{\theta} = -\varepsilon_\ell \tfrac{\pi}{T} a^2 a_\ell \mathsf{F}(\tfrac{3}{2}, 1) \sin(\phi_{\ell o} + \psi) \theta_{\ell o} \tag{4.44e}$$

Note that all right hand sides are functions of only *ratios* of semimajor axes, as $\mathsf{F}(\frac{3}{2}, \ldots)$ contains, in its definition, division by either a^3 or by a_ℓ^3. This implies that the solar system, expanded or deflated by a constant factor, would behave exactly the same as the existing one.

4.2.2 Decoupling and linearization

At this point we must realize that, in reality, there is not just one perturbed planet, but that *each* planet of the solar system is both being perturbed by, as well as perturbing, the remaining planets. This means that we have to simultaneously solve not just one set of four differential equations (4.44b)-(4.44e), but nine times as many, all coupled to each other.

4.2.2.1 Eccentricity and aphelion

To be able to solve the complete system (of the 36 differential equations), we first convert all Euler's angles to one common inertial frame, which must be chosen in such a way that all inclinations (relative to the x-y plane) remain small.

In this frame, $\cos\theta \simeq 1$, and the value of $\phi_{\ell o} + \psi_{\ell o}$ can be approximated by $\phi_\ell + \psi_\ell - \phi - \psi$ (this is easy to visualize). Equations (4.44b) and(4.44d) can thus be re-written as

$$\dot{\beta}_m = \sum_{\ell \neq m} H_{m,\ell} \sin(\phi_\ell + \psi_\ell - \phi_m - \psi_m) \beta_\ell \tag{4.45a}$$

$$\left(\dot{\psi}_m + \dot{\phi}_m\right) \beta_m = \beta_m \sum_{\ell \neq m} G_{m,\ell} - \sum_{\ell \neq m} H_{m,\ell} \cos(\phi_\ell + \psi_\ell - \phi_m - \psi_m) \beta_\ell \tag{4.45b}$$

where we have

1. started subscripting the perturbed-planet quantities by m (since now we have more than one),

2. made the summation over the remaining planets explicit,

3. and introduced the following shorthand notation

$$G_{m,\ell} = \varepsilon_\ell \tfrac{\pi}{T_m} a_m^2 a_\ell \mathsf{F}(\tfrac{3}{2}, 1) \tag{4.46a}$$

$$H_{m,\ell} = \varepsilon_\ell \tfrac{\pi}{T_m} a_m^2 a_\ell \mathsf{F}(\tfrac{3}{2}, 2) \tag{4.46b}$$

After introducing a new set of dependent variables (each one, being *complex*, stands for two of the old variables), namely

$$\mathcal{A}_j = \beta_j e^{\mathrm{i}(\phi_j+\psi_j)} \tag{4.47}$$

we can show that

$$\dot{\mathcal{A}}_m = \dot{\beta}_m e^{\mathrm{i}(\phi_m+\psi_m)} + \mathrm{i}\beta_m(\dot{\phi}_m + \dot{\psi}_m)e^{\mathrm{i}(\phi_m+\psi_m)} = \mathrm{i}\mathcal{A}_m \sum_{\ell\neq m} G_{m,\ell} - \mathrm{i}\sum_{\ell\neq m} H_{m,\ell}\mathcal{A}_\ell \tag{4.48}$$

which is now a *linear* set of differential equations, and can be thus written in the following matrix form

$$\dot{\mathbf{A}} = \mathrm{i}\mathbb{M}\mathbf{A} \tag{4.49}$$

where

$$\mathbb{M}_{m,\ell} = -H_{m,\ell} \quad \text{when } \ell \neq m \tag{4.50a}$$

$$\mathbb{M}_{m,m} = \sum_{\ell\neq m} G_{m,\ell} \quad \text{otherwise} \tag{4.50b}$$

To give a clear indication of who are the main perturbers of each planet, we present the following table of $H_{m,\ell}$ values (in mega-year^{-1}):

	M	V	E	A	J	S	U	N	P
M	-	13.88	4.63	0.12	7.87	0.37	0.01	0	0
V	0.69	-	35.59	0.43	20.40	0.96	0.02	0.01	0
E	0.16	24.67	-	1.25	34.31	1.57	0.03	0.01	0
A	0.03	2.23	9.42	-	71.20	3.04	0.05	0.02	0
J	0	0.02	0.05	0.01	-	35.45	0.39	0.11	0
S	0	0	0.01	0	87.35	-	1.44	0.32	0
U	0	0	0	0	4.45	6.65	-	2.00	0
N	0	0	0	0	0.84	0.98	1.35	-	0
P	0	0	0	0	0.32	0.35	0.32	2.99	-

(4.51)

(planets are identified by their first letters, with the exception of Mars, represented by A; one would find a similar pattern for $G_{m,\ell}$). Note that the strongest perturbers are usually the neighboring planets (one of them often dominating) and, in many cases, also Jupiter, as the biggest planet of the Solar system. This frequently creates pairs (such as Venus/Earth, Jupiter/Saturn and Uranus/Neptune) of two planets mirroring each other's behavior, e.g. their orbits' oscillations proceed in a synchronous, but opposite, manner.

To find a solution of (4.49), all we have to do is to compute eigenvalues (denoted λ_ℓ) and eigenvectors ($\mathbf{w}_\ell$) of $\mathbb{M}$, and combine them in the usual

$$\mathbf{A}(t) = \sum_j c_j \mathbf{w}_j e^{\mathrm{i}\lambda_j t} \tag{4.52}$$

fashion, where the c_j's are chosen to yield the correct set of initial values of each $\beta e^{\mathrm{i}(\phi+\psi)}$. We will not spell out the full solution (consisting of 9 eigenvalues and 81 complex coefficients), but only the lengths of the corresponding cycles (i.e. of $2\pi/\lambda_\ell$): 58.4, 72.1, 75.0, 177, 238, 351, 486, 1575 and 2082 millennia. Let us just mention that, to convert each component $\mathcal{A}_m(t)$ to the corresponding β_m and $\phi_m + \psi_m$, one only needs to take its absolute value and argument, respectively.

It is thus always the dominant (in terms of the overall coefficient of $e^{\mathrm{i}\lambda_j t}$) frequency which drives the advance of a planet's aphelion - the other frequencies only make this advance somehow irregular. For

Mercury, the main frequency is that of the 238 millennium cycle, resulting in a aphelion *average* advance of about 545″ per century (yet, it can range from less than 500″ to over 600″). The current value (from our computation) is 544″.1 per century (coincidentally, very close to the average value), missing the observed value of 531″.5 by about 2%, due to our approximations (note that we could easily remove these, to get a full agreement).

For Earth, the issue is somehow more complicated, as there are at least two dominant frequencies (with cycles of 72.1 and 177 millennia), and either of them can be the driving frequency for a time. It appears that the average time for Earth's perihelion to go a full circle is 130 millennia. This, combined with similar (but retrograde) motion of equinoxes (with a cycle of 25.4 millennia) results in a periodic change in Earth's climate (opposite for the two hemispheres), with a period of $(1/25.4 + 1/130)^{-1} = 21.2$ millennia. It is due to the fact that, when the closest approach to Sun happens in summer, seasonal effects get amplified (while, if the same happens during winters, these get mitigated). This is the first of such climatic variations, and is usually referred to as the PRECESSION CYCLE [26].

In terms of a planet's eccentricity, the predominant term makes its value constant; the other terms then cause more or less regular deviations from it; the period of any such deviation is computed based on the difference between the two (dominant and 'other') frequencies. For Earth, the main variation of its eccentricity is due to the difference between the second and the sixth frequency, having the period of

$$\left(\frac{1}{72.1} - \frac{1}{351}\right)^{-1} = 90.7 \text{ millennia} \tag{4.53}$$

This leads to the second, so called ECCENTRICITY CYCLE, also affecting Earth's climate, even though in a different way. It is due to the fact that the amount of solar energy received is inversely proportional to the distance from Sun squared (as radiation intensity decreases in this manner). The *average* yearly amount of energy is thus proportional to

$$\frac{1}{2\pi}\int_0^{2\pi} \frac{d\omega}{1 + \frac{2\beta}{1+\beta^2}\cos\omega} = \frac{1+\beta^2}{1-\beta^2} \tag{4.54}$$

i.e. it increases with eccentricity.

4.2.2.2 Orbit's normal

To convert equations (4.44c) and (4.44e) to an inertial frame is more difficult, as the right hand sides of (4.5b) and (4.5c) cannot be easily expanded in term of small θ and θ_ℓ (incidentally, their *sum* has a simple expansion, which was utilized earlier). To bypass this problem, we first visualize the two equations in Kepler's frame of the *perturbing* planet (implying that $\phi = -\psi_{\ell o}$, $\theta = -\theta_{\ell o}$ and $\psi = -\phi$ - we have to reverse signs and the order of Euler angles).

We then get

$$\dot{\phi} = -\varepsilon_\ell \tfrac{\pi}{T} a^2 a_\ell \mathsf{F}(\tfrac{3}{2}, 1) \tag{4.55a}$$

$$\dot{\theta} = 0 \tag{4.55b}$$

which can be expressed in an *frame-independent* manner of

$$\dot{\mathbf{u}} = -\varepsilon_\ell \tfrac{\pi}{T} a^2 a_\ell \mathsf{F}(\tfrac{3}{2}, 1)\mathbf{u}_\ell \times \mathbf{u} \tag{4.56}$$

where $\mathbf{u}$ and $\mathbf{u}_\ell$ are the *unit normals* of the orbits of the perturbed and unperturbed planets, respectively.

In an inertial frame (let us again use the current ecliptic), the x and y components of the previous equation read

$$\dot{\mathbf{u}}_x = -\varepsilon_\ell \tfrac{\pi}{T} a^2 a_\ell \mathsf{F}(\tfrac{3}{2}, 1)(\mathbf{u}_\ell - \mathbf{u})_y \tag{4.57a}$$

$$\dot{\mathbf{u}}_y = \varepsilon_\ell \tfrac{\pi}{T} a^2 a_\ell \mathsf{F}(\tfrac{3}{2}, 1)(\mathbf{u}_\ell - \mathbf{u})_x \tag{4.57b}$$

since both z-components are, to a good approximation, equal to 1. Introducing $\mathcal{B} = \mathbf{u}_x + i\ \mathbf{u}_y \simeq -\mathrm{i}\theta e^{\mathrm{i}\phi} = \theta e^{\mathrm{i}(\phi-\pi/2)}$ - see (1.22) - we can re-write this as:

$$\dot{\mathcal{B}} = \varepsilon_\ell \tfrac{\pi}{T} a^2 a_\ell \mathsf{F}(\tfrac{3}{2}, 1)\mathrm{i}(\mathcal{B}_\ell - \mathcal{B}) \tag{4.58}$$

Adding summation over all perturbing planets (and the m subscript to the perturbed planet), the last equation reads

$$\dot{\mathcal{B}}_m = -\mathrm{i}\mathcal{B}_m \sum_{\ell \neq m} G_{m,\ell} + \mathrm{i} \sum_{\ell \neq m} G_{m,\ell}\mathcal{B}_\ell \tag{4.59}$$

or, in matrix from

$$\dot{\mathbf{B}} = -\mathrm{i}\mathbb{N}\mathbf{B} \tag{4.60}$$

where

$$\mathbb{N}_{m,\ell} = -G_{m,\ell} \quad \text{when } \ell \neq m \tag{4.61a}$$

$$\mathbb{N}_{m,m} = \sum_{\ell \neq m} G_{m,\ell} \quad \text{otherwise} \tag{4.61b}$$

The solution will thus have the same for as (4.52), namely

$$\mathbf{B}(t) = \sum_j c_j \mathbf{w}_j e^{-\mathrm{i}\lambda_j t} \tag{4.62}$$

where λ_j and $\mathbf{w}_j$ are now the eigenvalues and eigenvectors of $\mathbb{N}$, and the c_j's are chosen to yield the correct set of initial values of $\theta e^{\mathrm{i}(\phi-\pi/2)}$. The resulting cycle lengths are: $50.6, 69.4, 73.6, 197, 250, 453, 1575$ and 1946 millennia; this time we get only eight of them, since one of the eigenvalues (say the last one) is always equal to zero ($\mathbb{N}$ is clearly singular, as its columns add up to $\mathbf{0}$). The corresponding contribution of the $\lambda_9 = 0$ eigenvalue to (4.62), namely $c_9\mathbf{w}_9$, has all components equal to $0.0262 + 0.0083\mathrm{i}$. This translates into x and y coordinates of the *common axis*, around which all $\mathbf{u}$-vectors (including the ecliptic's normal) circulate. For Earth, this circulation is rather irregular (having a period of about 69.4 millennia), as shown in Figure 4.1. (Note that the solution has started at the origin, due to our choice of inertial frame — the current ecliptic.)

This yields the third periodic variation of Earth's climate (the so called OBLIQUITY CYCLE), even though now the situation is somehow more complicated. The effect is due to Earth's *axis* (whose unit direction we denote $\mathbf{v}$) precessing around $\mathbf{u}$ (Earth orbit's normal) with a period of about 25.4 millennia. This would not normally change Earth's obliquity (the angle between the two directions, which currently stands at 23°27′), if it were not for the fact that $\mathbf{u}$ itself precesses in the manner just described. For simplicity, we assume the latter precession to be perfectly regular, i.e.

$$\mathbf{u} = \exp(\frac{\mathbf{\Omega}}{2}t) \circ \mathbf{u}_0 \circ \exp(-\frac{\mathbf{\Omega}}{2}t) \tag{4.63}$$

where $|\mathbf{\Omega}| = 2\pi/69.4/\text{millennia}$, and the direction of $\mathbf{\Omega}$ is that of the common-axis: $\langle 0.0262, 0.0083, 0.9996\rangle$, using our favourite inertial frame (in which $\mathbf{u}_0$ is a unit vector along z).

For $\mathbf{v}$, we thus get the following differential equation

$$\dot{\mathbf{v}} = -g\mathbf{u} \times \mathbf{v} \tag{4.64}$$

where $g \simeq 2\pi/25.4$ millennia^{-1}. Note that *both* precessions are *retrograde*.

It is easy to verify that the solution to (4.64) is

$$\mathbf{v} = \exp(\tfrac{\mathbf{\Omega}}{2}t) \circ \exp(\tfrac{g\mathbf{u}_0 - \mathbf{\Omega}}{2}t) \circ \mathbf{v}_0 \circ \exp(-\tfrac{g\mathbf{u}_0 - \mathbf{\Omega}}{2}t) \circ \exp(-\tfrac{\mathbf{\Omega}}{2}t) \tag{4.65}$$

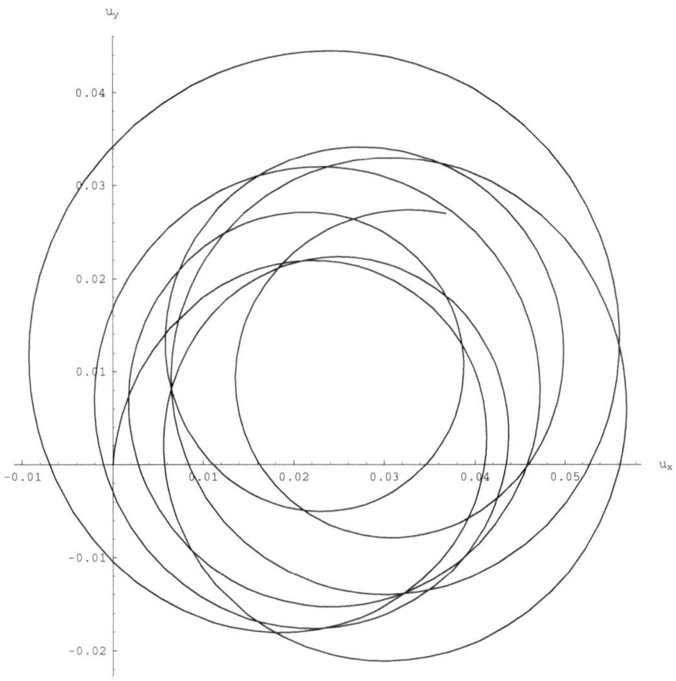

Figure 4.1 Earth Orbit's Normal.

Proof.

$$\begin{aligned}
\dot{\mathbf{v}} &= \tfrac{\mathbf{\Omega}}{2} \circ \mathbf{v} - \mathbf{v} \circ \tfrac{\mathbf{\Omega}}{2} \\
&\quad + \exp(\tfrac{\mathbf{\Omega}}{2}t) \circ \tfrac{g\mathbf{u}_0 - \mathbf{\Omega}}{2} \circ \exp(-\tfrac{\mathbf{\Omega}}{2}t) \circ \exp(\tfrac{\mathbf{\Omega}}{2}t) \circ \exp(\tfrac{g\mathbf{u}_0 - \mathbf{\Omega}}{2}t) \circ \mathbf{v}_0 \circ \exp(-\tfrac{g\mathbf{u}_0 - \mathbf{\Omega}}{2}t) \circ \exp(-\tfrac{\mathbf{\Omega}}{2}t) \\
&\quad - \exp(\tfrac{\mathbf{\Omega}}{2}t) \circ \exp(\tfrac{g\mathbf{u}_0 - \mathbf{\Omega}}{2}t) \circ \mathbf{v}_0 \circ \exp(-\tfrac{g\mathbf{u}_0 - \mathbf{\Omega}}{2}t) \circ \exp(-\tfrac{\mathbf{\Omega}}{2}t) \circ \exp(\tfrac{\mathbf{\Omega}}{2}t) \circ \tfrac{g\mathbf{u}_0 - \mathbf{\Omega}}{2} \circ \exp(-\tfrac{\mathbf{\Omega}}{2}t) \\
&= -\mathbf{\Omega} \times \mathbf{v} + \tfrac{g\mathbf{u} - \mathbf{\Omega}}{2} \circ \mathbf{v} - \mathbf{v} \circ \tfrac{g\mathbf{u} - \mathbf{\Omega}}{2} = -\mathbf{\Omega} \times \mathbf{v} - (g\mathbf{u} - \mathbf{\Omega}) \times \mathbf{v} = -g\mathbf{u} \times \qquad (4.66)
\end{aligned}$$

□

The corresponding obliquity is the angle between $\mathbf{u}$ and $\mathbf{v}$, or equivalently between $\exp(-\mathbf{\Omega}t/2) \circ \mathbf{u} \circ \exp(\mathbf{\Omega}t/2)$ and $\exp(-\mathbf{\Omega}t/2) \circ \mathbf{v} \circ \exp(\mathbf{\Omega}t/2)$, i.e. between $\mathbf{u}_0$ and

$$\exp(\tfrac{g\mathbf{u}_0 - \mathbf{\Omega}}{2}t) \circ \mathbf{v}_0 \circ \exp(-\tfrac{g\mathbf{u}_0 - \mathbf{\Omega}}{2}t) \qquad (4.67)$$

where $\mathbf{v}_0$ is at an angle of $23^o 27'$ to the z axis, and has a zero x-coordinate (the direction of vernal equinox). Clearly, (4.67) rotates around the direction of $g\mathbf{u}_0 - \mathbf{\Omega}$ (which tilts by an angle of 0.91^o from z) correspondingly increasing and decreasing its angle with $\mathbf{u}_0$ (which is parallel to z). Note that the actual variation of this angle (Earth's obliqueness) is somehow higher than $\pm 0.91^o$, due to the irregular precession of $\mathbf{u}$; one can get correct value ($\simeq \pm 1.2^o$) by numerically integrating (4.64). Nevertheless, the corresponding angular speed of this precession is quite regular, and it equals to $|g\mathbf{u}_0 - \mathbf{\Omega}| \simeq g - \Omega$. Based on this, Earth's obliquity experiences a variation with the period of $2\pi/(g - \Omega)$, which translates to $(1/25.4 - 1/69.4)^{-1} = 40.1$ millennia.

TWO PLANET SOLUTION:

Finally, let us have a closer look at a phenomenon (mentioned in the previous section) of two neighboring planets having such strong influence on each other, that their orbital planes precess in unison, as if firmly joined together.

In this context, it is interesting to solve the hypothetical case of a system with only two planets, namely

$$\dot{\mathbf{u}}_1 = -G_{1,2}\mathbf{u}_2 \times \mathbf{u}_1 \tag{4.68a}$$

$$\dot{\mathbf{u}}_2 = -G_{2,1}\mathbf{u}_1 \times \mathbf{u}_2 \tag{4.68b}$$

The two equations have a rather simple solution, given by

$$\mathbf{u}_1 = \exp(-\frac{\mathbf{\Omega}}{2}t) \circ \mathbf{a}_1 \circ \exp(\frac{\mathbf{\Omega}}{2}t) \tag{4.69a}$$

$$\mathbf{u}_2 = \exp(-\frac{\mathbf{\Omega}}{2}t) \circ \mathbf{a}_2 \circ \exp(\frac{\mathbf{\Omega}}{2}t) \tag{4.69b}$$

where $\mathbf{a}_1$ and $\mathbf{a}_2$ are arbitrary (constant) *unit* vectors, and $\mathbf{\Omega} = -G_{2,1}\mathbf{a}_1 - G_{1,2}\mathbf{a}_2$ (the negative signs implying that the precession is retrograde). This means that the two normals precess around a common axes $\mathbf{\Omega}$, with the same angular speed of $G_{1,2} + G_{2,1}$ (using the small-angle approximation), the individual inclinations (from $\mathbf{\Omega}$) being proportional to the respective values of $G_{1,2}$ and $G_{2,1}$.

Proof.

$$\begin{aligned}
\dot{\mathbf{u}}_1 &= \exp(-\frac{\mathbf{\Omega}}{2}t) \circ \left(-\frac{\mathbf{\Omega}}{2} \circ \mathbf{a}_1 + \mathbf{a}_1 \circ \frac{\mathbf{\Omega}}{2}\right) \circ \exp(\frac{\mathbf{\Omega}}{2}t) \\
&= \exp(-\frac{\mathbf{\Omega}}{2}t) \circ (\mathbf{\Omega} \times \mathbf{a}_1) \circ \exp(\frac{\mathbf{\Omega}}{2}t) \\
&= -G_{1,2} \exp(-\frac{\mathbf{\Omega}}{2}t) \circ (\mathbf{a}_2 \times \mathbf{a}_1) \circ \exp(\frac{\mathbf{\Omega}}{2}t) \\
&= -G_{1,2} \exp(-\frac{\mathbf{\Omega}}{2}t) \circ \frac{\mathbf{a}_1\mathbf{a}_2 - \mathbf{a}_2\mathbf{a}_1}{2} \circ \exp(\frac{\mathbf{\Omega}}{2}t) \\
&= -G_{1,2}\frac{\mathbf{u}_1\mathbf{u}_2 - \mathbf{u}_2\mathbf{u}_1}{2} = -G_{1,2}\mathbf{u}_2 \times \mathbf{u}_1
\end{aligned} \tag{4.70}$$

and the same, after the $1 \leftrightarrow 2$ interchange.

Also: $\mathbf{a}_1 \times \mathbf{\Omega} = -G_{1,2}(\mathbf{a}_1 \times \mathbf{a}_2)$ and $\mathbf{a}_2 \times \mathbf{\Omega} = G_{2,1}(\mathbf{a}_1 \times \mathbf{a}_2)$. For nearly parallel $\mathbf{a}_1$ and $\mathbf{a}_2$, the inclination of the two orbits is proportional to $|\mathbf{a}_1 \times \mathbf{\Omega}|$ and $|\mathbf{a}_2 \times \mathbf{\Omega}|$ respectively, and is thus in the $G_{1,2} \div G_{2,1}$ ratio. □

For the Jupiter-Saturn pair, this yields a fairly adequate solution, with a period of 51.2 millennia (instead of 50.6 millennia we have obtained using all planets).

CHAPTER 5

Oblateness perturbations

Abstract

The fact that Earth is not a perfect sphere (it is slightly flattened and otherwise distorted, as discussed in a previous chapter), introduces extra terms into the expansion of its gravitational potential, which in turn affect the motion of near-Earth (i.e. artificial) satellites. The first one of these terms (due to Earth's OBLATENESS) is by far the largest, and therefore quite often the only one considered (finding the corresponding solution is referred to as the MAIN PROBLEM of satellite motion)[1]. At the same time, it is the quintessential example of time-independent perturbations, and the main topic of this chapter.

The remaining terms are of two basic types: those corresponding to axially symmetric (e.g. pear shaped) distortions are called ZONAL HARMONICS, and remain time-independent; those which vary with longitude (and thus *rotate* with Earth) are called TESSERAL HARMONICS. For these, we only mention a few basic results.

5.1 PERTURBING FORCE

As we have seen already (3.26), the perturbing force due to Earth's at-the-poles flattening is

$$\varepsilon \mathbf{f} = -\frac{\varepsilon M}{r^5}\left(\mathbf{r} - 5\frac{(\mathbf{r}\cdot\mathbf{u})^2\mathbf{r}}{r^2} + 2(\mathbf{r}\cdot\mathbf{u})\mathbf{u}\right) \tag{5.1}$$

where $\varepsilon = (3/2)J_2 R_\ominus^2$ (atypically, *not* dimensionless), and $\mathbf{u}$ is the direction of Earth's axis. Since the satellites' mass is negligible, we can identify M with μ.

If we choose an inertial system in which $\mathbf{u} = \mathfrak{i}$ then, in the satellite's Kepler frame (used for all vector quantities from now on)

$$\mathbf{u}_\mathrm{o} = \mathbb{R} \circ \mathfrak{i} \circ \overline{\mathbb{R}} = \mathfrak{i}\cos\theta + \mathfrak{j}\sin\theta\exp(-\mathrm{i}\psi) \tag{5.2}$$

(see Eq. 1.17). Since

$$\mathbf{r}_\mathrm{o} = \mathfrak{k}\frac{az(1+\frac{\beta}{z})^2}{(1+\beta^2)} \tag{5.3}$$

(see Eq. 1.59) and

$$r = \frac{a(1+\frac{\beta}{z})(1+\beta z)}{(1+\beta^2)} \tag{5.4}$$

(see Eq. 1.60), one can easily compute

$$\frac{\mathbf{r}_\mathrm{o}\cdot\mathbf{u}_\mathrm{o}}{r} = \mathrm{i}\frac{\sin\theta}{2}\left(\frac{1+\beta z}{z+\beta}\exp(-\mathrm{i}\psi) - \frac{z+\beta}{1+\beta z}\exp(\mathrm{i}\psi)\right) \tag{5.5}$$

[1]For more details see: [6], [15], [22] and [44].

and

$$\frac{\mathbf{f}_\text{o}}{\varepsilon\mu} = \frac{(1+\beta^2)^4 z^4}{8a^4}\left(5\frac{\mathfrak{k}[\cos(2\theta)-1]\exp(2\mathrm{i}\psi)}{(z+\beta)(1+\beta z)^7} - \frac{4\sin(2\theta)\exp(\mathrm{i}\psi)+\mathfrak{k}[6\cos(2\theta)+2]}{(z+\beta)^3(1+\beta z)^5}\right.$$
$$\left.+\frac{4\sin(2\theta)\exp(-\mathrm{i}\psi)+\mathfrak{k}[\cos(2\theta)-1]\exp(-2\mathrm{i}\psi)}{(z+\beta)^5(1+\beta z)^3}\right) \tag{5.6}$$

5.2 FIRST-ORDER RESULTS

As the first step, we construct the ε-accurate solution (dropping the corresponding subscript, as we can easily keep track of our iterations)

The previous expression implies that

$$\frac{\mathcal{Q}(z)}{\varepsilon} = -\frac{2a\mathrm{Cx}(\mathbf{r}_\text{o}\mathbf{f}_\text{o})}{\varepsilon\mu(1+\beta z)} = \frac{(1+\beta^2)^3 z^3}{4a^2}\left(5\frac{[\cos(2\theta)-1]\exp(2\mathrm{i}\psi)}{(z+\beta)(1+\beta z)^6}\right.$$
$$\left.-\frac{[6\cos(2\theta)+2]}{(z+\beta)^3(1+\beta z)^4}+\frac{[\cos(2\theta)-1]\exp(-2\mathrm{i}\psi)}{(z+\beta)^5(1+\beta z)^2}\right) \tag{5.7}$$

and

$$\frac{\mathcal{W}(z)}{\varepsilon} = -\frac{4ar\mathrm{Cx}\left(\mathbf{f}_\text{o}\right)}{\varepsilon\mu} = \frac{2(1+\beta^2)^3 z^3\sin(2\theta)}{a^2}\left(\frac{\exp(\mathrm{i}\psi)}{(z+\beta)^2(1+\beta z)^4} - \frac{\exp(-\mathrm{i}\psi)}{(z+\beta)^4(1+\beta z)^2}\right) \tag{5.8}$$

With the help of the residue theorem (the only singularities inside the unit disk are at $z=-\beta$ and, occasionally, at $z=0$), we get, based on the following contour-integral version of (2.29a):

$$a' = 2a\mathrm{Im}\oint_{C_0}(1-\beta z)\mathcal{Q}(z)\frac{dz}{2\pi\mathrm{i}z} = 0 \tag{5.9}$$

since

$$\oint_{C_0}(1-\beta z)\frac{z^3}{(z+\beta)(1+\beta z)^6}\frac{dz}{2\pi\mathrm{i}z} = \frac{\beta^2(1+\beta^2)}{(1-\beta^2)^6} \tag{5.10a}$$

$$\oint_{C_0}(1-\beta z)\frac{z^3}{(z+\beta)^3(1+\beta z)^4}\frac{dz}{2\pi\mathrm{i}z} = \frac{1+9\beta^2+9\beta^4+\beta^6}{(1-\beta^2)^6} \tag{5.10b}$$

and

$$\oint_{C_0}(1-\beta z)\frac{z^3}{(z+\beta)^5(1+\beta z)^2}\frac{dz}{2\pi\mathrm{i}z} = \frac{5\beta^2(1+\beta^2)}{(1-\beta^2)^6} \tag{5.11}$$

Similarly, we obtain

$$\beta' = -\frac{1+\beta^2}{4}\mathrm{Im}\oint_{C_0}(\frac{1}{z}+3\beta+3z+\beta z^2)\mathcal{Q}(z)\frac{dz}{2\pi\mathrm{i}z} = 0 \tag{5.12}$$

$$Z_1 = -\frac{1}{1-\beta^2}\mathrm{Im}\oint_{C_0}\left(\frac{1+\beta^2}{2z}+\beta\right)\mathcal{W}(z)\frac{dz}{2\pi\mathrm{i}z} = -\frac{\varepsilon}{a^2}\frac{(1+\beta^2)^4}{(1-\beta^2)^4}\sin 2\theta\sin\psi \tag{5.13}$$

$$Z_2 = -\frac{1}{2}\mathrm{Re}\oint_{C_0}\frac{1}{z}\mathcal{W}(z)\frac{dz}{2\pi \mathrm{i} z} = -\frac{\varepsilon}{a^2}\frac{(1+\beta^2)^4}{(1-\beta^2)^4}\sin 2\theta\cos\psi \tag{5.14}$$

$$Z_3 = \frac{1}{4\beta}\mathrm{Re}\oint_{C_0}\left(-\frac{1+\beta^2}{z}+(1-3\beta^2)\beta+(3-\beta^2)z+(1+\beta^2)\beta z^2\right)\mathcal{Q}(z)\frac{dz}{2\pi \mathrm{i} z} = \frac{\varepsilon}{a^2}\frac{(1+\beta^2)^4}{(1-\beta^2)^4}\frac{1+3\cos 2\theta}{2} \tag{5.15}$$

The last three expressions can be easily converted (see Eqs. (1.19)) to:

$$\phi' = -\frac{2\varepsilon}{a^2}\frac{(1+\beta^2)^4}{(1-\beta^2)^4}\cos\theta \tag{5.16a}$$

$$\theta' = 0 \tag{5.16b}$$

$$\psi' = \frac{\varepsilon}{a^2}\frac{(1+\beta^2)^4}{(1-\beta^2)^4}(4-5\sin^2\theta) \tag{5.16c}$$

The first of these indicates that the nodes experience a steady retrograde precession, whose rate decreases with increasing inclination. The last one yields a similar precession of the line of apses (connecting perigee and apogee), *relative* to the nodes. This time, the precession is retrograde when $\theta < 63.435^{\mathrm{o}}$ (the so called CRITICAL INCLINATION) and direct otherwise.

To complete the ε-accurate solution, we still need (these are computed in analogous manner)

$$s_{\mathrm{p}}' = \frac{\varepsilon}{4a^2}\frac{(1+\beta^2)^3}{(1-\beta^2)^3}(2-3\sin^2\theta) \tag{5.17a}$$

$$b = -\frac{3\varepsilon}{4a^2}\frac{(1+\beta^2)^3}{(1-\beta^2)^3}\beta\sin 2\theta\sin\psi \tag{5.17b}$$

$$\frac{a^2}{\varepsilon}\mathcal{D}(z) = \frac{(1+\beta^2)^3(2+\beta z)z^2(1-\cos 2\theta)}{48(1+\beta z)^3}\exp(2\mathrm{i}\psi) + \frac{(1+\beta^2)^3(1+3\cos 2\theta)}{24(1-\beta^2)^4 z}\times \left[\beta\frac{-2(1+2\beta^2)+2\beta(1-\beta^2)z+(-1+6\beta^2+\beta^4)z^2}{1+\beta z} - 2(2\beta+\beta^3+z+2\beta^2 z)\log\frac{1+\beta z}{1+\frac{\beta}{z}}\right] \tag{5.17c}$$

and

$$\frac{a^2}{\varepsilon}\mathcal{S}(z) = \frac{\mathrm{i}(1+\beta^2)^4\sin 2\theta}{4(1-\beta^2)^4}\mathrm{Im}\left\{\exp(\mathrm{i}\psi)\left[z(1+\frac{\beta}{z})^2\log\frac{1+\beta z}{1+\frac{\beta}{z}}+\frac{3\beta^2}{2z}+\beta(1-\beta^2) + \frac{z}{1+\beta^2}\left(\frac{2}{3}-4\beta^2+\frac{1}{2}\beta^4-\frac{1}{6}\beta^6-\frac{2(1-\beta^2)^3}{3(1+\beta z)}\right)\right]\right\} \tag{5.17d}$$

The last three quantities translate into the following additional terms to $\mathbf{r}_{\mathrm{o}}$:

$$\mathbf{r}_{\mathrm{o}} = \cdots + \mathfrak{k}\frac{\varepsilon\sin^2\theta z^3}{6a}\exp(2\mathrm{i}\psi) - \mathfrak{k}\frac{\varepsilon\beta(1+3\cos 2\theta)z^2}{4a} + \mathfrak{k}\frac{\varepsilon\beta\sin^2\theta(2z^2-5z^4)}{12a}\exp(2\mathrm{i}\psi) + \mathrm{i}\frac{\varepsilon\beta\sin 2\theta}{6a}\mathrm{Im}\left[(9-5z^2)\exp(\mathrm{i}\psi)\right] + O(\beta^2) \tag{5.18}$$

where the three dots represent the zero-order part (5.3). To simplify the last result, we have expanded it in β, keeping the first two terms only.

Note that the first of these additional terms corresponds to 'flattening' (reducing the radius of) the orbit at Earth's poles by $J_2 R_\ominus^2/(4a)$; to see that, re-write $z^3\exp(2\mathrm{i}\psi)$ as

$$\exp[3\mathrm{i}(\omega+\psi)]\exp(-\mathrm{i}\psi) \tag{5.19}$$

where $\omega+\psi$ is the satellite's angular distance from the equator, and the second factor rotates the result back to Kepler's frame.

For non-polar orbits, this effect decreases with $\sin^2\theta$, similarly to Earth's own shape. But even for polar orbits close to Earth, the relative size of this flattening is about an order of magnitude smaller than that of Earth's: $J_2/4$, as compared to $5J_2/2$ of (3.32).

Similarly, r (5.4) acquires the extra

$$r=\cdots+\frac{\varepsilon\sin^2\theta}{6a}\mathrm{Re}\left[z^2(1-\tfrac{3}{2}\beta z)\exp(2\mathrm{i}\psi)\right]-\frac{\varepsilon\beta(1+3\cos 2\theta)}{8a}(z+\tfrac{1}{z})+O(\beta^2) \tag{5.20}$$

Note that all differentiation is done with respect to the modified time s. Usually, we like to convert d../ds to d../dω, where ω is the physically more meaningful eccentric anomaly, via a simple division by $2(1-s_\mathrm{p}')$. In this case (of oblateness perturbations), even more convenient (as an independent variable) is $\omega+\psi$, since its increase by 2π represents one *nodal* orbit; the corresponding conversion would divide all derivatives by $2(1-s_\mathrm{p}')+\psi'$. For ε-accurate answers, the two approaches are identical (a division by 2 is sufficient in either case).

5.3 SECOND-ORDER CORRECTIONS

We will now investigate in more detail the subtleties of our solution at the critical inclination ($\theta\simeq 63.435^\mathrm{o}$). To do that, we have to extend our solution to ε^2-accuracy. To simplify computation, we now expand all results in powers of β, quoting only their lowest *nonzero* terms.

We will again drop the second-iteration subscript; we have to remember that these results are to be *added* to those of the first iteration (indicated by three dots). Here they are:

$$a'=0$$

$$\beta'=-\frac{\varepsilon^2\beta}{12a^4}\sin 2\psi\sin^2\theta(11+15\cos 2\theta) \tag{5.21a}$$

$$\phi'=\cdots+\frac{\varepsilon^2}{12a^4}\cos\theta(5+19\cos 2\theta) \tag{5.21b}$$

$$\theta'=\frac{\varepsilon^2\beta^2}{3a^4}\sin 2\psi\,(19\sin 2\theta+15\sin 4\theta) \tag{5.21c}$$

$$\psi'=\cdots+\frac{\varepsilon^2}{a^4}\left[-2+\frac{19}{6}\sin^2\theta-\frac{55}{24}\sin^4\theta+\left(\frac{13}{3}\sin^2\theta-5\sin^4\theta\right)\sin^2\psi\right] \tag{5.21d}$$

$$s_\mathrm{p}'=\cdots-\frac{\varepsilon^2}{144a^4}(9-35\cos 2\theta+8\cos 4\theta)\sin^2\theta-\frac{\varepsilon^2}{24a^4}(11+15\cos 2\theta)\sin^2\theta\cos 2\psi \tag{5.21e}$$

and

$$\begin{aligned}\mathbf{r}_\mathrm{o}=\cdots&-\mathfrak{k}\frac{\varepsilon^2(11+15\cos 2\theta)\sin^2\theta}{96a^3 z}\exp(-2\mathrm{i}\psi)-\mathfrak{k}\frac{\varepsilon^2(51+79\cos 2\theta)\sin^2\theta}{288a^3}z^3\exp(2\mathrm{i}\psi)\\&+\mathfrak{k}\frac{\varepsilon^2\sin^4\theta}{144a^3}z^5\exp(4\mathrm{i}\psi)-\mathfrak{i}\frac{\varepsilon^2\cos\theta\sin^3\theta}{8a^3}\mathrm{Im}[z^3\exp(3\mathrm{i}\psi)]\end{aligned} \tag{5.21f}$$

We can see that a is still, to this approximation, a constant (it would acquire a small, oscillating term only in the next iteration). Unlike a, β and θ have become (rather slowly varying) functions of time, and our objective now is to provide a solution to the three differential equations for β', θ' and ψ', at or near the critical value of θ (a traditional challenge, which many iterative techniques fail to treat adequately).

5.3.1 Critical inclination[2]

To investigate the solution to (5.21c) and (5.21d) in the vicinity of the critical angle, we may replace $\sin^2\theta$ by 4/5 (correspondingly, $\sin 2\theta$ by 4/5, $\sin 4\theta$ by $-24/25$ and $\cos 2\theta$ by $-3/5$) in all ε^2-proportional terms [55]. The equations further simplify if we introduce a new dependent variable

$$\Theta \equiv \frac{(1+\beta^2)^4}{(1-\beta^2)^4}\left(5\sin^2\theta - 4\right) \tag{5.22}$$

in place of the original θ. Now, they read

$$\beta' = -\frac{2\varepsilon^2\beta}{15a^4}\sin 2\psi \tag{5.23a}$$

$$\Theta' = \frac{16}{15}\frac{\varepsilon^2}{a^4}\beta^2\sin 2\psi \tag{5.23b}$$

(since, at this level, we are expanding in β and replacing $\sin^2\theta$ by 4/5), and

$$\psi' = -\frac{\varepsilon}{a^2}\Theta + \frac{\varepsilon^2}{a^4}\left(-\frac{14}{15} + \frac{4}{15}\sin^2\psi\right) \tag{5.23c}$$

It is immediately apparent that

$$K = \Theta + 4\beta^2 \tag{5.24}$$

is a (small) constant of motion (note that this implies $\Theta \geq K$). Eliminating β from (5.23b) results in

$$\Theta' = \frac{4}{15}\frac{\varepsilon^2}{a^4}(K-\Theta)\sin 2\psi \tag{5.25}$$

Equations (5.23c) and (5.25) have two fixed points at $\psi = 0$ (or π) and $\Theta = -14\varepsilon/(15a^2)$, and another two at $\psi = \pm\pi/2$ and $\Theta = -2\varepsilon/(3a^2)$. When $K < -14\varepsilon/(15a^2)$, none of these are in the 'physical' (i.e. $\Theta \leq K$) region; when $-14\varepsilon/(15a^2) < K < -2\varepsilon/(3a^2)$ the second pair is still out, but the first one ($\psi = 0$ or π) represents two disconnected (i.e. separated by circulating solutions) elliptic CENTERS of librating solutions. Finally, when $K > -2\varepsilon/(3a^2)$, basins of the two centers become adjacent, separated only by the remaining ($\psi = \pm\pi/2$) fixed points, which are of the hyperbolic (SADDLE-POINT) type.

The linearized version of (5.23c) and (5.25) around the two centers reads

$$\Psi' = -\frac{\varepsilon}{a^2}\left(\Theta + \frac{14}{15}\frac{\varepsilon}{a^2}\right) \tag{5.26a}$$

$$\left(\Theta + \frac{14}{15}\frac{\varepsilon}{a^2}\right)' = \frac{8}{15}\frac{\varepsilon^2}{a^4}\left(K + \frac{14}{15}\frac{\varepsilon}{a^2}\right)\Psi \tag{5.26b}$$

(where $\Psi = \psi$ or $\Psi = \psi - \pi$) and the corresponding solution is

$$\Psi = A\cos(\lambda s - \delta) \tag{5.27a}$$

$$\Theta = -\frac{14}{15}\frac{\varepsilon}{a^2} + \sqrt{\frac{8\varepsilon\left(K + \frac{14}{15}\frac{\varepsilon}{a^2}\right)}{15a^2}}\,A\sin(\lambda s - \delta) \tag{5.27b}$$

[2] Compare with [10].

where

$$\lambda = \frac{\varepsilon}{a^2}\sqrt{\frac{8\varepsilon\left(K + \frac{14}{15}\frac{\varepsilon}{a^2}\right)}{15a^2}} \tag{5.28}$$

and A and δ are arbitrary constants. Note that one libration cycle is completed in $2/\lambda$ satellite's orbits.

Exact solutions of (5.23c) and (5.25) can be constructed based on the fact that

$$\left(\Theta + \frac{4}{5}\frac{\varepsilon}{a^2}\right)^2 + \frac{4}{15}\frac{\varepsilon}{a^2}(\Theta - K)\cos 2\psi \tag{5.29}$$

is also a constant of motion, as can be easily verified (differentiate the above expression, substitute Θ' and ψ' from the two equations, and cancel all terms). Plotting the contours of this function (of two arguments: $\Theta \simeq 5\sin^2\theta - 4$ and ψ), using $\varepsilon/a^2 = 0.0015$ and $K = 13\varepsilon/(15a^2)$, enables us to illustrate a whole family of potential solutions, as done in Figure 5.1.

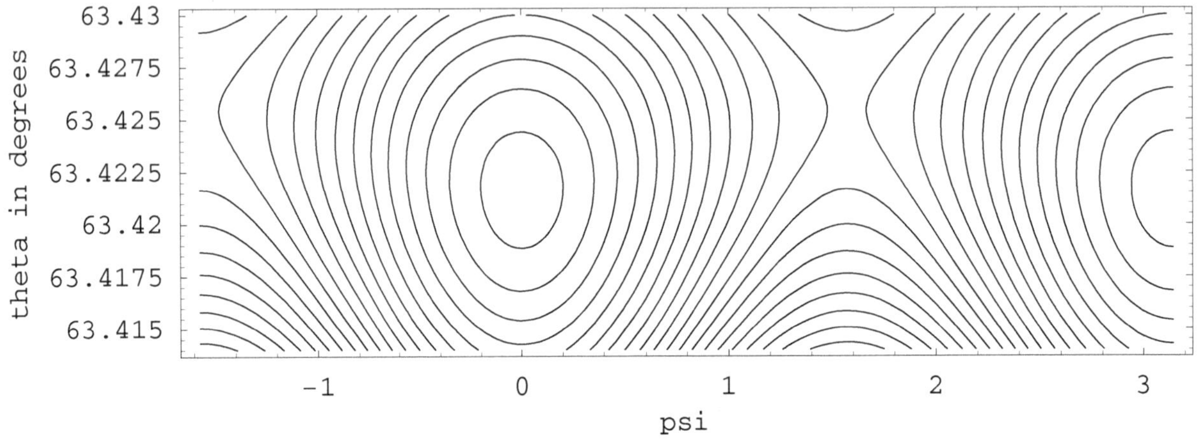

Figure 5.1 Librating solutions at critical inclination.

In view of (5.22), one can slightly improve the approximation by using

$$\Theta \simeq (5\sin^2\theta - 4)\left(1 - 2K + 2(5\sin^2\theta - 4)\right)$$

instead; in the above graph, this would have been hardly noticeable.

So, the only new detail which emerges at or near the critical inclination (compared to first-order solution) is a creation of small pockets of librating (in terms of ψ) solutions. These are always centered at $\psi = 0$ or π and $\theta \simeq 63.4225°$ (slightly below the critical value), and become larger (in terms of their θ extent) as β increases.

5.4 HIGHER ZONAL HARMONICS[3]

Defying the chapter's title, we now quickly consider terms of the (3.22) expansion with $n > 2$, to the first order of accuracy. This time, we define dimensionless

$$\varepsilon_n = J_n\left(\frac{R_\ominus}{a}\right)^n \tag{5.30}$$

[3]SEE [53].

expand all expressions in β, and quote only the leading terms of the resulting corrections (without any computational details)

$$a' = 0 \tag{5.31a}$$

$$\begin{aligned}\beta' &= \varepsilon_3 \frac{3(\sin\theta + 5\sin 3\theta)}{32}\cos\psi - \varepsilon_4 \frac{15\beta(5 + 7\cos 2\theta)\sin^2\theta}{32}\sin 2\psi \\ &\quad - \varepsilon_5 \frac{15(2\sin\theta + 7\sin 3\theta + 21\sin 5\theta)}{512}\cos\psi + \cdots\end{aligned} \tag{5.31b}$$

$$\begin{aligned}\phi' &= -\varepsilon_3 \frac{3\beta(\cos\theta + 15\cos 3\theta)}{8\sin\theta}\sin\psi + \varepsilon_4 \frac{15(9\cos\theta + 7\cos 3\theta)}{32} \\ &\quad + \varepsilon_5 \frac{15\beta(2\cos\theta + 21\cos 3\theta + 104\cos 5\theta)}{128\sin\theta}\sin\psi + \cdots\end{aligned} \tag{5.31c}$$

$$\begin{aligned}\theta' &= -\varepsilon_3 \frac{3\beta(11\cos\theta + 5\cos 3\theta)}{8}\cos\psi + \varepsilon_4 \frac{15\beta^2(10\sin 2\theta + 7\sin 4\theta)}{32}\sin 2\psi \\ &\quad + \varepsilon_5 \frac{15\beta(58\cos\theta + 49\cos 3\theta + 21\cos 5\theta)}{128}\cos\psi + \cdots\end{aligned} \tag{5.31d}$$

$$\begin{aligned}\psi' &= -\varepsilon_3 \frac{3(3 + 5\cos 2\theta)\sin\theta}{16\beta}\sin\psi - \varepsilon_4 \frac{15(27 + 52\cos 2\theta + 49\cos 4\theta)}{128} \\ &\quad - \varepsilon_4 \frac{15(5 + 7\cos 2\theta)\sin^2\theta}{32}\cos 2\psi + \varepsilon_5 \frac{15(15 + 28\cos 2\theta + 21\cos 4\theta)\sin\theta}{256\beta}\sin\psi + \cdots\end{aligned} \tag{5.31e}$$

$$\begin{aligned}\psi' - 2s'_{\mathrm{p}} &= -\varepsilon_3 \frac{3\beta(3 - 12\cos 2\theta - 55\cos 4\theta)}{32\sin\theta}\sin\psi - \varepsilon_4 \frac{15(81 + 148\cos 2\theta + 91\cos 4\theta)}{512} \\ &\quad + \varepsilon_5 \frac{15\beta(10 - 11\cos 2\theta - 154\cos 4\theta - 357\cos 6\theta)}{512\sin\theta}\sin\psi + \cdots\end{aligned} \tag{5.31f}$$

$$\begin{aligned}\mathbf{r}_{\mathrm{o}} = &-\mathfrak{j}\frac{5\varepsilon_3 a\sin^3\theta}{32}z^4\exp(3\mathrm{i}\psi) - \mathfrak{j}\frac{3\varepsilon_3 a(\sin\theta + 5\sin 3\theta)}{32}z^2\exp(\mathrm{i}\psi) - \mathfrak{k}\frac{7\varepsilon_4 a\sin^4\theta}{64}z^5\exp(4\mathrm{i}\psi) \\ &-\mathfrak{k}\frac{5\varepsilon_4 a(5 + 7\cos 2\theta)\sin^2\theta}{64}z^3\exp(2\mathrm{i}\psi) - \mathfrak{k}\frac{5\varepsilon_4 a(5 + 7\cos 2\theta)\sin^2\theta}{64z}\exp(-2\mathrm{i}\psi) \\ &+\mathfrak{j}\frac{21\varepsilon_5 a\sin^5\theta}{256}z^6\exp(5\mathrm{i}\psi) + \mathfrak{j}\frac{35\varepsilon_5 a(7 + 9\cos 2\theta)\sin^3\theta}{768}z^4\exp(3\mathrm{i}\psi) \\ &+\mathfrak{j}\frac{45\varepsilon_5 a(15 + 28\cos 2\theta + 21\cos 4\theta)\sin\theta}{1024}z^2\exp(\mathrm{i}\psi) - \mathfrak{j}\frac{35\varepsilon_5 a(7 + 9\cos 2\theta)\sin^3\theta}{1536z^2}\exp(-3\mathrm{i}\psi) + \cdots \\ &-\mathfrak{i}\frac{3\varepsilon_3 a(1 - 5\cos 2\theta)\cos\theta}{8} - \mathfrak{i}\frac{5\varepsilon_3 a\cos\theta\sin^2\theta}{4}\mathrm{Re}[z^2\exp(2\mathrm{i}\psi)] - \mathfrak{i}\frac{35\varepsilon_4 a\cos\theta\sin^3\theta}{64}\mathrm{Im}[z^3\exp(3\mathrm{i}\psi)] \\ &-\mathfrak{i}\frac{15\varepsilon_5 a(29 - 28\cos 2\theta + 63\cos 4\theta)\cos\theta}{512} + \mathfrak{i}\frac{35\varepsilon_5 a(1 + 3\cos 2\theta)\cos\theta\sin^2\theta}{32}\mathrm{Re}[z^2\exp(2\mathrm{i}\psi)] \\ &+\mathfrak{i}\frac{21\varepsilon_5 a\cos\theta\sin^4\theta}{64}\mathrm{Re}[z^4\exp(4\mathrm{i}\psi)] + \cdots\end{aligned} \tag{5.31g}$$

These should be added to the (first and second order) terms of the oblateness ($n = 2$) perturbations, derived in the previous section.

5.5 SECTORIAL HARMONICS

And, finally and briefly, we consider perturbations caused by a primary which is *not* axially symmetric. Expanding its potential will thus lead to terms which depend on the location's longitude; those which do not depend on latitude are called SECTORIAL harmonics, those which depend on both latitude and longitude are called TESSERAL harmonics [58]. Here we discuss only the simplest example of the former, corresponding to $n = 2$ and $m = 2$ in (3.21), namely

$$V(r, \tilde{\theta}, \varphi_t) = -\frac{3J_{22}MR_{\ominus}^2(1 - \cos^2\tilde{\theta})}{r^3}\cos(2\varphi_t - 2\varphi_0) \tag{5.32}$$

where φ_0 is a constant which, for simplicity, we set equal to zero (which can always be done by the proper choice of x and y axes; the z axis has been aligned with the primary's polar direction). Note that the coordinate system is *rotating* with the primary, which we indicate by attaching t as a subscript to the corresponding coordinate(s).

In rectangular coordinates, this reads

$$V = -\varepsilon_{22}M\frac{(x_t^2 - y_t^2)}{r^5} \tag{5.33}$$

where $\varepsilon_{22} \equiv 3J_{22}R_{\ominus}^2$, since

$$x_t = r\sin\tilde{\theta}\cos\varphi_t \tag{5.34a}$$
$$y_t = r\sin\tilde{\theta}\sin\varphi_t \tag{5.34b}$$

Taking the gradient of V, we get the following perturbing force (using *regular* vector notation)

$$\varepsilon_{22}\mathbf{f} = 5\varepsilon_{22}M\frac{(x_t^2 - y_t^2)\mathbf{r}_t}{r^7} - \varepsilon_{22}M\frac{\langle 2x_t, -2y_t, 0\rangle}{r^5} \tag{5.35}$$

To express this force in an inertial frame (with the same z axis as the rotating frame), we take $\mathbf{m} = \mathfrak{k}\exp(\mathrm{i}\Omega_0 t)$ to be the direction of the *rotating* x axis (Ω_0 is thus the primary's rotational speed) and $\mathbf{n} = \mathbf{m}\circ\mathfrak{i}$ to be the direction of the rotating y axis (note that the two frames coincide at $t = 0$). This yields $x_t = \mathbf{r}\cdot\mathbf{m}$ and $y_t = \mathbf{r}\cdot\mathbf{n}$, which enables us to express the perturbing force thus:

$$\varepsilon_{22}\mathbf{f} = 5\varepsilon_{22}M\frac{(\mathbf{r}\cdot\mathbf{m})^2 - (\mathbf{r}\cdot\mathbf{n})^2}{r^7}\mathbf{r} - 2\varepsilon_{22}M\frac{(\mathbf{r}\cdot\mathbf{m})\mathbf{m} - (\mathbf{r}\cdot\mathbf{n})\mathbf{n}}{r^5} \tag{5.36}$$

The rest is a routine application of time-dependent formulas of Chapter 2, resulting in (quoting only the leading term of each β expansion)

$$\frac{\mathrm{d}a}{\mathrm{d}\omega} = \frac{\varepsilon_{22}}{a}\frac{3\Omega\sin^2\theta}{1-4\Omega^2}\sin(2\phi - 2\Omega\omega) \tag{5.37a}$$

$$\frac{\mathrm{d}\beta}{\mathrm{d}\omega} = \frac{\varepsilon_{22}}{a^2}\beta\left(\frac{\Omega^2\sin^4\frac{\theta}{2}}{(1+2\Omega)^2}\sin(2\phi - 2\psi - 2\Omega\omega) - \frac{\Omega^2\cos^4\frac{\theta}{2}}{(1-2\Omega)^2}\sin(2\phi + 2\psi - 2\Omega\omega)\right.$$
$$\left. - \frac{3\Omega(1-8\Omega^2)\sin^2\theta}{(1-4\Omega^2)^2}\sin(2\phi - 2\Omega\omega)\right) \tag{5.37b}$$

$$\frac{\mathrm{d}\phi}{\mathrm{d}\omega} = -\frac{\varepsilon_{22}}{a^2}\frac{\cos\theta + \Omega}{1-\Omega^2}\cos(2\phi - 2\Omega\omega) \tag{5.37c}$$

$$\frac{\mathrm{d}\theta}{\mathrm{d}\omega} = -\frac{\varepsilon_{22}}{a^2}\frac{(1+\Omega\cos\theta)\sin\theta}{1-\Omega^2}\sin(2\phi - 2\Omega\omega) \tag{5.37d}$$

$$\frac{\mathrm{d}\psi}{\mathrm{d}\omega} - 2\frac{\mathrm{d}s_{\mathrm{p}}}{\mathrm{d}\omega} = \frac{\varepsilon_{22}}{a^2}\left(\frac{(\cos\theta + \Omega)\cos\theta}{1-\Omega^2} - \frac{3\sin^2\theta}{4(1-4\Omega^2)}\right)\cos(2\phi - 2\Omega\omega) \tag{5.37e}$$

$$\frac{\mathrm{d}s_{\mathrm{p}}}{\mathrm{d}\omega} = -\frac{\varepsilon_{22}}{2a^2}\left(\frac{\Omega^2\sin^4\frac{\theta}{2}}{(1+2\Omega)^2}\cos(2\phi - 2\psi - 2\Omega\omega) + \frac{\Omega^2\cos^4\frac{\theta}{2}}{(1-2\Omega)^2}\cos(2\phi + 2\psi - 2\Omega\omega)\right.$$
$$\left. + \frac{3(1-8\Omega^2-16\Omega^4)\sin^2\theta}{4(1-4\Omega^2)^2}\cos(2\phi - 2\Omega\omega)\right) \tag{5.37f}$$

and

$$\mathbf{r}_{\mathrm{o}} = \cdots - \mathfrak{k}\frac{\varepsilon_{22}}{4a}\left(\frac{\Omega\sin^4\frac{\theta}{2}\exp[2\mathfrak{i}(\phi - \Omega\omega - \psi)]}{(1+\Omega)^2(1+2\Omega)z} + \frac{(4-5\Omega)\cos^4\frac{\theta}{2}\exp[2\mathfrak{i}(\phi - \Omega\omega + \psi)]}{(1-\Omega)^2(3-2\Omega)}z^3\right.$$
$$\left.\frac{(4+5\Omega)\sin^4\frac{\theta}{2}\exp[-2\mathfrak{i}(\phi - \Omega\omega - \psi)]}{(1+\Omega)^2(3+2\Omega)}z^3 - \frac{\Omega\cos^4\frac{\theta}{2}\exp[-2\mathfrak{i}(\phi - \Omega\omega + \psi)]}{(1-\Omega)^2(1-2\Omega)z}\right)$$
$$+\mathfrak{i}\frac{\varepsilon_{22}}{a}\beta\sin\theta\left(\frac{(3+10\Omega+4\Omega^2)\sin^2\frac{\theta}{2}\sin(2\phi - 2\Omega\omega - \psi)}{(1+\Omega)(1-4\Omega^2)} + \frac{(3-10\Omega+4\Omega^2)\cos^2\frac{\theta}{2}\sin(2\phi - 2\Omega\omega + \psi)}{(1-\Omega)(1-4\Omega^2)}\right.$$
$$\left.-\frac{(5+2\Omega-4\Omega^2)\sin^2\frac{\theta}{2}\sin(2\phi - 2\Omega\omega - \psi - 2\omega)}{(1+\Omega)(1+2\Omega)(3+2\Omega)} - \frac{(5-2\Omega-4\Omega^2)\sin^2\frac{\theta}{2}\sin(2\phi - 2\Omega\omega + \psi + 2\omega)}{(1-\Omega)(1-2\Omega)(3-2\Omega)}\right) \tag{5.38}$$

where, in the last expression, three dots represent the unperturbed solution, and

$$\Omega = \Omega_0\sqrt{\frac{a_{\mathrm{o}}^3}{M}} \tag{5.39}$$

which equals to the satellite's orbital period, relative to the length of the perturbing-force cycle; for close-to-Earth artificial satellites, Ω has a value of about 0.059.

Note that, since we work in the first-order-of-ε_{22} accuracy, $\Omega\omega$ is to a sufficient approximation the angle of the rotating x axis (the old $\Omega_0 t$), and that the d../ds to d../dω conversion has been achieved by a simple division by 2.

It would be pointless to solve these differential equations, as they never tell the full story, and serve only as minor corrections to other perturbations (oblateness, etc.).

It is interesting to note (which is the main point of this section) that the above formulas become singular at $\Omega = 1/2$, $\Omega = 1$ and $\Omega = 3/2$, and are therefore valid only for small (say $\Omega < 0.5$) values of Ω (this can be verified by comparing, numerically, our solution with the exact one). If one needs a good solution in the $0.5 \leq \Omega < 0.75$ range, it is necessary to replace Ω by $1/2 + \delta$ (where δ is 'small'), and construct a new solution (using the same time-dependent method of Chapter 2, which *automatically* adjusts itself to the new situation — this is discussed in Section 2.3.1).

Now, we get

$$\frac{\mathrm{d}a}{\mathrm{d}\omega} = -\frac{\varepsilon_{22}}{a}\beta\left(\frac{3(1+\delta)\sin^2\theta}{1-2\delta}\sin(2\phi - 2\delta\omega) - \frac{2(1-\delta-6\delta^2)\cos^4\frac{\theta}{2}}{(1-2\delta)^2}\sin(2\phi + 2\psi - 2\delta\omega)\right) \tag{5.40}$$

which is quite different from the previous solution (but equally accurate in the intended range of Ω values, as one can verify by numerical simulation).

To simplify the remaining results, we quote them assuming that $\Omega = 1/2$, i.e. for the exact 2/1 resonance (a satellite with a period equal to half of sidereal day).

$$\frac{\mathrm{d}\beta}{\mathrm{d}\omega} = -\frac{\varepsilon_{22}}{a^2}\left(\frac{3\sin^2\theta}{8}\sin 2\phi + \frac{\cos^4\frac{\theta}{2}}{4}\sin(2\phi + 2\psi)\right) \tag{5.41a}$$

$$\frac{\mathrm{d}\phi}{\mathrm{d}\omega} = \frac{\varepsilon_{22}}{a^2}\beta\left(3\cos\theta\cos 2\phi + \frac{1+\cos\theta}{2}\cos(2\phi + 2\psi)\right) \tag{5.41b}$$

$$\frac{\mathrm{d}\theta}{\mathrm{d}\omega} = \frac{\varepsilon_{22}}{a^2}\beta\left(3\sin\theta\sin 2\phi - \frac{(1+\cos\theta)\sin\theta}{2}\sin(2\phi + 2\psi)\right) \tag{5.41c}$$

$$\frac{\mathrm{d}\psi}{\mathrm{d}\omega} - 2\frac{\mathrm{d}s_{\mathrm{p}}}{\mathrm{d}\omega} = -\frac{\varepsilon_{22}}{a^2}\beta\left(3\cos 2\theta\cos 2\phi + (1+\cos\theta)\cos^2\frac{\theta}{2}\sin(2\phi + 2\psi)\right) \tag{5.41d}$$

$$\frac{\mathrm{d}s_{\mathrm{p}}}{\mathrm{d}\omega} = \frac{\varepsilon_{22}}{16a^2\beta}\left(3\sin^2\theta\cos 2\phi - 2\cos^4\frac{\theta}{2}\cos(2\phi + 2\psi)\right) \tag{5.41e}$$

and

$$\mathbf{r}_{\mathrm{o}} = \cdots - \mathfrak{t}\frac{\varepsilon_{22}}{a}\left(\frac{3z^2}{4}\cos^4\frac{\theta}{2}\exp[2\mathrm{i}(\phi+\psi)] + \frac{1}{36z^2}\sin^4\frac{\theta}{2} + \frac{13z^4}{72}\sin^4\frac{\theta}{2}\exp[2\mathrm{i}(-\phi+\psi)]\right.$$
$$\left. + \frac{3z^2}{8}\sin^2\theta\exp(-2\mathrm{i}\phi)\right) - \mathrm{i}\frac{\varepsilon_{22}}{a}\sin\theta\left((1+\cos\theta)\sin(2\phi+\psi) - \frac{1-\cos\theta}{3}\sin(2\phi - \psi - 2\omega)\right) \tag{5.42}$$

CHAPTER 6

Lunar problem

Abstract

In this chapter we investigate complexities of Moon's motion [11], [49], [53]. The perturbing body is Sun itself; the magnitude of the perturbing force is thus relatively large. Here, we can no longer use the averaging principle (its error would skew some of our results by as much as a factor of two), thus, the time-dependent version of our formulas will be occasionally required. Traditionally, this problem has always been considered a benchmark for any new technique.

6.1 PERTURBING FORCE

We will start by considering only the first two terms of (3.17), the second one being the actual perturbing force

$$\varepsilon \mathbf{f} \equiv -\frac{\mu \varepsilon A^3}{a_{\mathrm{o}}^3 R^3}\left(\mathbf{r} - 3(\mathbf{r}\cdot\mathbf{R})\frac{\mathbf{R}}{R^2}\right) \tag{6.1}$$

where μ is the total mass of the Earth-Moon system, a_{o} is the average length of Moon's semimajor axis,

$$\varepsilon \equiv \frac{M}{\mu}\frac{a_{\mathrm{o}}^3}{A^3} \simeq 0.005580 \tag{6.2}$$

(M and A are Sun's mass and semimajor axis, respectively), and $\mathbf{R}$ is the Sun's location relative to the Earth' center. Now, all locations are expressed in this manner; we are thus using a geocentric system in which not only Moon but also Sun is seen as rotating around Earth.

To simplify the computation, we expand all expressions in terms of three other small parameters, namely $\beta \simeq 0.02747$ and $\theta \simeq 0.08959$ (Moon's average modified eccentricity and inclination, respectively - the latter in radians) and $\gamma \simeq 0.01673$ (Sun's *ordinary* eccentricity). Normally, we consider only their first-order contributions.

To this accuracy, $\mathbf{R}$ equals to (combining (1.65) and (1.63), and replacing e by $-\gamma$, a by A, and τ by $\Omega_0 t$, where $\Omega_0 \equiv \sqrt{M/A^3} \simeq 2\pi/365.266\ \text{day}^{-1}$)

$$\mathbf{R} = A\mathfrak{k}\left(\exp(\mathrm{i}\Omega_0 t) - \frac{3\gamma}{2} + \frac{\gamma}{2}\exp(2\mathrm{i}\Omega_0 t)\right) + O(\gamma^2) \tag{6.3}$$

using (1.64) with $e \to -\gamma$ (we let $t = 0$ occur at Earth's perihelion). This readily implies that

$$R = A\left(1 - \gamma\cos(\Omega_0 t)\right) + O(\gamma^2) \tag{6.4}$$

The unperturbed Moon's orbit can be similarly expanded (Eq. 1.59) as follows

$$\mathbf{r} = \overline{\mathbb{R}}\mathfrak{k}a(z + 2\beta)\mathbb{R} + O(\beta^2) \tag{6.5}$$

where, by utilizing (1.17),

$$\mathbb{R} = \exp(\mathrm{i}\frac{\phi+\psi}{2}) + \mathfrak{k}\frac{\theta}{2}\exp(\mathrm{i}\frac{\phi-\psi}{2}) + O(\theta^2) \tag{6.6}$$

Jan Vrbik

Similarly

$$r = a\left(1 + \beta(z + \frac{1}{z})\right) + O(\beta^2) \tag{6.7}$$

by (1.60), and

$$\mathbb{U} = \sqrt{a}(q + \beta q^{-1})\mathbb{R} + O(\beta^2) \tag{6.8}$$

according to (1.105).

Based on (1.61), we also obtain, to the ε^0 accuracy

$$t = 2\int \sqrt{\frac{a}{\mu}} r \, \mathrm{d}s = \int \sqrt{\frac{a}{\mu}} \frac{r}{1 - s_p'} \mathrm{d}\omega \simeq \sqrt{\frac{a_{\mathrm{o}}^3}{\mu}} \left(\omega - \omega_0 - i\beta(z - \frac{1}{z})\right) + O(\beta^2) \tag{6.9}$$

where ω_0 is a constant (the value of ω at $t = 0$, to a good approximation).

In Moon's Kepler frame we get (all results quoted to the first-order in β, θ and γ accuracy)

$$\mathbf{r}_{\mathrm{o}} = \mathfrak{k} a(z + 2\beta) \tag{6.10}$$

and, with the help of (6.9)

$$\begin{aligned} \mathbf{R}_{\mathrm{o}} &\equiv \mathbb{R}\mathbf{R}\overline{\mathbb{R}} \\ &= A\mathrm{i}\theta \sin[-(\Omega_0 t - \phi)] + A\mathfrak{k}\left(\exp(\mathrm{i}\Omega_0 t) - \frac{3\gamma}{2} + \frac{\gamma}{2}\exp(2\mathrm{i}\Omega_0 t)\right)\exp[-\mathrm{i}(\phi + \psi)] \\ &= A\frac{\theta}{2}\left(\frac{\exp(\mathrm{i}\phi)}{u} - u\exp(-\mathrm{i}\phi)\right) + A\mathfrak{k}\left(u + \beta\Omega(z - \frac{1}{z})u - \frac{3\gamma}{2} + \frac{\gamma}{2}u^2\right)\exp[-\mathrm{i}(\phi + \psi)] \end{aligned} \tag{6.11}$$

where $\Omega \equiv \sqrt{a_{\mathrm{o}}^3/\mu}\Omega_0 \simeq \sqrt{\varepsilon}$ (duration of anomalistic month, divided by sidereal year), and $u \equiv e^{-\mathrm{i}\Omega\omega_0} z^{\Omega}$.

Similarly,

$$R = A\left(1 - \frac{\gamma}{2}(u + \frac{1}{u})\right) \tag{6.12}$$

Based on these formulas, we can now easily expand $(4a/\mu)\mathbf{f}_{\mathrm{o}}$, getting, to the same approximation:

$$\begin{aligned} &\frac{1}{a}\mathfrak{k}\left[2z + 4\beta + u^2\left(\frac{6}{z} + 12\beta(1 + \Omega) - \frac{12\beta\Omega}{z^2}\right)\exp[-2\mathrm{i}(\phi + \psi)] + 3\gamma z(u + \frac{1}{u})\right. \\ &\left. + \frac{3\gamma}{z}(7u^3 - u)\exp[-2\mathrm{i}(\phi + \psi)]\right] \\ &+ 3\frac{\theta}{a}\left(\frac{\exp(-\mathrm{i}\psi)}{z} - z\exp(\mathrm{i}\psi) + \frac{z}{u^2}\exp[\mathrm{i}(2\phi + \psi)] - \frac{u^2}{z}\exp[-\mathrm{i}(2\phi + \psi)]\right) \end{aligned} \tag{6.13}$$

This further implies that

$$\mathcal{Q}(z) \equiv -2\frac{a}{\mu}\varepsilon\frac{\mathrm{Cx}(\mathbf{r}_{\mathrm{o}}\mathbf{f}_{\mathrm{o}})}{1 + \beta z} \tag{6.14}$$

can be expanded to yield

$$\begin{aligned} &\varepsilon\left[1 + \beta(z + \frac{2}{z}) + 3u^2\left(\frac{1}{z^2} + \frac{(3 + 2\Omega)\beta}{z} - \frac{2\beta\Omega}{z^3}\right)\exp[-2\mathrm{i}(\phi + \psi)] + \frac{3\gamma}{2}(u + \frac{1}{u})\right. \\ &\left. + \frac{3\gamma(7u^3 - u)}{2z^2}\exp[-2\mathrm{i}(\phi + \psi)]\right] \end{aligned} \tag{6.15}$$

and, similarly

$$\mathcal{W}(z) \equiv -4\frac{a}{\mu}\varepsilon r \mathrm{Cx}(\mathbf{f}_\mathrm{o}) \tag{6.16}$$

results in

$$3\varepsilon\theta\left(z\exp(\mathrm{i}\psi) - \frac{\exp(-\mathrm{i}\psi)}{z} - \frac{z}{u^2}\exp[\mathrm{i}(2\phi+\psi)] + \frac{u^2}{z}\exp[-\mathrm{i}(2\phi+\psi)]\right) \tag{6.17}$$

6.2 FIRST-ORDER SOLUTION

Formulas (2.57) now enable us to construct the corresponding ε-accurate solution. Here, we present the individual results. They have been converted from the usual d../ds derivatives (denoted by a prime) to the more convenient derivatives with respect to ϖ, where $\varpi \equiv \omega - \omega_0$, via a simple division by $2(1 - s'_\mathrm{p}) \simeq 2$.

For the rate of change of semi-major axis, we get

$$\frac{\mathrm{d}a}{\mathrm{d}\varpi} = -\frac{3}{2}\varepsilon a \frac{\gamma\Omega}{1-\Omega^2}\sin(\Omega\varpi) \tag{6.18}$$

which integrates to

$$\ln\frac{a}{a_\mathrm{o}} = \frac{3}{2}\frac{\varepsilon\gamma}{1-\Omega^2}\cos(\Omega\varpi) \simeq \frac{3}{2}\frac{\varepsilon\gamma}{1-\Omega^2}\cos(\Omega_0 t) \tag{6.19}$$

or, at this level of approximation,

$$a = a_\mathrm{o}\left(1 + \frac{3}{2}\frac{\varepsilon\gamma}{1-\Omega^2}\cos(\Omega_0 t)\right) \tag{6.20}$$

yielding a $\pm 0.014\%$ yearly variation in the value of a. The largest value of a is observed at Earth's perihelion.

Similarly,

$$\frac{\mathrm{d}\beta}{\mathrm{d}\varpi} = -\varepsilon\beta\frac{3(5 - 5\Omega - 4\Omega^2)}{4(1-2\Omega)^2}\sin(2\Omega\varpi - 2\phi - 2\psi) \tag{6.21}$$

which integrates to

$$\beta = \beta_\mathrm{o}\left(1 + \varepsilon\frac{3(5 - 5\Omega - 4\Omega^2)}{4(1-2\Omega)^2}\frac{\cos(2\Omega_0 t - 2\phi - 2\psi)}{2\Omega}\right) \tag{6.22}$$

as ϕ and ψ can be considered constant (their rate of change is an order of magnitude smaller than that of $\Omega_0 t$).

When substituting $\Omega = 0.07544$ (using the empirical, correct value of Ω, instead of the approximate $\Omega \simeq \sqrt{\varepsilon} \cong 0.07470$), this yields

$$\beta = \beta_\mathrm{o}\left(1 + 0.177\cos(2\Omega_0 t - 2\phi - 2\psi)\right) \tag{6.23}$$

implying a fairly substantial periodic variation of Moon's eccentricity. Note that $\Omega t - \phi - \psi$ corresponds to Sun's longitude, *relative* to Moon's apogee, and that β is the largest when Sun's direction becomes aligned with Moon's line of apses.

6.2.1 Nodal precession

The next equation we can extract from $\mathcal{Q}(z)$ is

$$\frac{\mathrm{d}\phi}{\mathrm{d}\varpi} = -\frac{3}{4}\varepsilon\left(1 - \frac{\cos(2\Omega\varpi - 2\phi)}{1-\Omega}\right) \tag{6.24}$$

This time, it is possible to solve it exactly, which we first do this in its fully general form of

$$y' = -A + B\cos(Cx - Dy) \tag{6.25}$$

where A, B, C and D are four parameters, and x and y is the independent and dependent variable, respectively.

This last equation is equivalent to

$$C - Dy' = C + AD - BD\cos(Cx - Dy) \tag{6.26}$$

or

$$Y' = a - b\cos Y \tag{6.27}$$

where

$$Y = Cx - Dy \tag{6.28a}$$
$$b = BD \tag{6.28b}$$

and

$$a = C + AD \tag{6.29}$$

One can easily verify that (6.26) has the following solution

$$\begin{aligned} Y - Y_0 &= 2\arctan\left(\sqrt{\frac{a-b}{a+b}}\tan\left[\tfrac{\sqrt{a^2-b^2}}{2}x\right]\right) \\ &= \sqrt{a^2-b^2}\,x + 2\sum_{n=1}^{\infty}\left(\frac{\sqrt{\frac{a-b}{a+b}}-1}{\sqrt{\frac{a-b}{a+b}}+1}\right)^n \frac{\sin\left(nx\sqrt{a^2-b^2}\right)}{n} \end{aligned} \tag{6.30}$$

Note that the multivalued arctan(..) is capable of keeping the solution *continuous* as x increases, which is clearly required in this case. The last equality follows from the following Fourier expansion

$$\arctan(\alpha\tan\xi) - \xi = \sum_{n=1}^{\infty}\left(\frac{\alpha-1}{\alpha+1}\right)^n \frac{\sin(2n\xi)}{n} \tag{6.31}$$

valid for all ξ, provided that $\arctan(\alpha\tan\xi)$ is continuous, making $\arctan(\alpha\tan\xi) - \xi$ a periodic function of ξ (whose period is equal to π).

This implies that the *secular* part of the solution to (6.25) is

$$y_{\text{sec}} = \left(\tfrac{C}{D} - \sqrt{\left(\tfrac{C}{D}+A\right)^2 - B^2}\right)x + \cdots \tag{6.32}$$

the three dots implying periodic terms, of which the largest one is

$$y_{\rm osc} = -\frac{2}{D}\frac{\sqrt{\frac{C}{D}+A-B}-\sqrt{\frac{C}{D}+A+B}}{\sqrt{\frac{C}{D}+A-B}+\sqrt{\frac{C}{D}+A+B}} \sin\left[\left(\sqrt{(C/D+A)^2-B^2}\right)Dx\right]+\cdots$$

$$= -\frac{2}{D}\frac{\sqrt{\frac{C}{D}+A-B}-\sqrt{\frac{C}{D}+A+B}}{\sqrt{\frac{C}{D}+A-B}+\sqrt{\frac{C}{D}+A+B}} \sin\left[Cx - D(y_{\rm sec}-y_0)\right]+\cdots \tag{6.33}$$

Returning now to the original equation, namely (6.24), we thus get, for the SECULAR part of the solution

$$\phi_{\rm sec} = \left(\Omega - \sqrt{\left(\Omega+\frac{3}{4}\varepsilon\right)^2 - \left(\frac{\frac{3}{4}\varepsilon}{1-\Omega}\right)^2}\right)\varpi + \cdots$$

$$= \left(1 - \sqrt{\left(1+\frac{3\Omega}{4}\right)^2 - \left(\frac{3\Omega}{4(1-\Omega)}\right)^2}\right)\Omega_0 t + \cdots \tag{6.34}$$

and for the leading oscillatory term (the coefficient has been expanded in ε)

$$\phi_{\rm osc} = -\frac{3\varepsilon}{8\Omega(1-\Omega)}\left(1 - \frac{3\varepsilon}{4\Omega}+\cdots\right)\sin\left[2\Omega_0 t - 2\phi_{\rm sec}\right] \tag{6.35}$$

Including the initial value, the solution then reads

$$\phi = \phi_0 + \phi_{\rm sec} + \phi_{\rm osc} + \cdots \tag{6.36}$$

The reciprocal of the big parentheses in (6.34) yields the period of one NODAL CYCLE (in years, since $2\pi/\Omega_0 = 1$ year). It evaluates to -18.60 (the negative sign implying that the precession is retrograde), in excellent (even though somehow fortuitous) agreement with the observed value of 18.61 years. Note that this precession is the fastest (about double its average speed) when Sun is aligned with one of the two nodes, the slowest (down to zero, even reversing its direction for a while) when Sun's direction is perpendicular to the nodal direction.

Also note that, when (6.24) is solved using the averaging principle (which makes the cos term disappear), one gets, for the same period, a less accurate value of

$$\frac{\Omega}{-\frac{3\varepsilon}{4}} \simeq -17.85 \text{ years} \tag{6.37}$$

When the computation is extended to include terms quadratic and cubic (none found) in β, θ and γ, the of right hand side of (6.24) acquires an extra

$$-\varepsilon(6\beta^2+\frac{9}{8}\gamma^2-\frac{3}{8}\theta^2)+\varepsilon\left(\frac{6(1-\Omega+\Omega^3)}{(1-\Omega)(1-2\Omega)}\beta^2 - \frac{15\gamma^2}{8(1-\Omega)} - \frac{3\theta^2}{8(1-\Omega^2)}\right)\cos(2\Omega\varpi-2\phi) \tag{6.38}$$

the dots now representing 'dissonant' oscillating terms, i.e. those proportional to $\cos(\Omega\varpi)$, $\cos(\Omega\varpi-2\phi)$, $\cos(2\psi)$, etc. - they contribute only minute amounts to the secular part of the solution.

Adding (6.38) to (6.24) yields, for the length of one nodal cycle, a 'corrected' value of 18.55 years. Further corrections would arise due to higher-order-in-ε terms.

6.2.2 Perigee precession

Similarly, for the precession of the LINE OF APSES, we get the following differential equation, also of type (6.25):

$$\frac{d(\phi+\psi)}{d\varpi} = \frac{3}{4}\varepsilon\left(1 + \frac{5-5\Omega-4\Omega^2}{(1-2\Omega)^2}\cos(2\Omega\varpi - 2\phi - 2\psi)\right) \tag{6.39}$$

The exact solution yields the following secular term

$$\begin{aligned}(\phi+\psi)_{\text{sec}} &= \left(\Omega - \sqrt{(\Omega - \frac{3}{4}\varepsilon)^2 - (\frac{3}{4}\varepsilon)^2\left(\frac{5-5\Omega-4\Omega^2}{(1-2\Omega)^2}\right)^2}\right)\varpi + \cdots \\ &\simeq \left(1 - \sqrt{\left(1-\frac{3\Omega}{4}\right)^2 - \frac{9\Omega^2(5-5\Omega-4\Omega^2)^2}{16(1-2\Omega)^4}}\right)\Omega_0 t + \cdots\end{aligned} \tag{6.40}$$

(note that the initial value of $\phi+\psi$ is 0, by our definition of ϖ and of the inertial frame), and the leading oscillatory term of

$$(\phi+\psi)_{\text{osc}} \simeq \frac{15\varepsilon}{8\Omega}\sin\left[2\Omega_0 t - 2(\phi+\psi)_{\text{sec}}\right] \tag{6.41}$$

The coefficient of the secular term corresponds to a precession cycle of 8.05 years (the observed value is 8.85 years). Note that, based on (6.39), the precession proceeds at three times its average speed when the Sun's direction is parallel with Moon's line of apses, it reverses its direction in the perpendicular case.

The averaging principle would result in a grossly inaccurate value of

$$\frac{\Omega}{\frac{3\varepsilon}{4}} \simeq 17.85 \text{ years} \tag{6.42}$$

which used to trouble Newton a great deal.

Note that, when only ψ_{sec} is considered (precession of apses relative to *nodal* direction), one gets, for the duration of one cycle, the value of

$$\left(\tfrac{1}{8.85} + \tfrac{1}{18.61}\right)^{-1} = 5.998 \text{ years} \tag{6.43}$$

which is in nearly exact 6/1 resonance with the perturbing force. This poses a slight challenge when constructing high-accuracy solutions (which goes beyond the scope of this book).

To improve the accuracy of our computation of the perigee's precession cycle, the right hand side of (6.39) can be extended by the following quadratic (in β, γ and θ) terms

$$\begin{aligned}&-\frac{3\varepsilon}{2}\left(\beta^2 - \frac{3}{4}\gamma^2 + \theta^2\right) \\ &-\frac{\varepsilon}{2}\beta^2\frac{15+45\Omega+27\Omega^2+22\Omega^3-64\Omega^4-152\Omega^5-48\Omega^6}{(1-4\Omega^2)^2}\cos(2\Omega\varpi-2\phi-2\psi) \\ &-\frac{3\varepsilon}{8}\left(5\gamma^2+\theta^2\right)\frac{5-5\Omega-4\Omega^2}{(1-2\Omega)^2}\cos(2\Omega\varpi-2\phi-2\psi)\end{aligned} \tag{6.44}$$

which results in only a slightly improved value of 8.17 years. More substantial corrections are due to higher-order-in-ε terms discussed below.

6.2.3 Nutation

For the rate of change of θ, we get

$$\frac{\mathrm{d}\theta}{\mathrm{d}\varpi} = -\frac{3}{4}\varepsilon\theta\frac{\sin(2\Omega\varpi - 2\phi)}{1-\Omega} \tag{6.45}$$

This, to a sufficient approximation (now, the constant-ϕ assumption is reasonable), integrates to

$$\theta = \theta_o\left(1 + \frac{3\varepsilon}{8(1-\Omega)\Omega}\cos(2\Omega_0 t - 2\phi)\right) \tag{6.46}$$

which corresponds to a $\pm 3.02\%$ variation in the value of θ. Note that $\Omega_0 t - \phi$ is Sun's longitude, relative to Moon's nodal direction.

6.2.4 Orbit's distortion

In the current approximation

$$b = 0 + \cdots \tag{6.47a}$$

$$\mathcal{S}(z) = 0 + \cdots \tag{6.47b}$$

and

$$\begin{aligned}\mathcal{D}(z) &= -\frac{3\varepsilon z^2}{32(1-\Omega)^2(3-2\Omega)}\exp[-\mathrm{i}(2\Omega\varpi - 2\phi - 2\psi)] \\ &\quad - \frac{3(5-4\Omega)\varepsilon}{32z^2(1-\Omega)^2(1-2\Omega)}\exp[\mathrm{i}(2\Omega\varpi - 2\phi - 2\psi)] + \cdots \\ &= -\frac{3\varepsilon z_\Lambda^2}{32(1-\Omega)^2(3-2\Omega)} - \frac{3(5-4\Omega)\varepsilon}{32z_\Lambda^2(1-\Omega)^2(1-2\Omega)} + \cdots\end{aligned} \tag{6.48}$$

where $z_\Lambda = \exp[\mathrm{i}(\omega - \Lambda)]$ and $\Lambda \equiv \Omega_0 t - \phi - \psi$ (Sun's longitude in Moon's Kepler frame), and three dots indicating terms proportional to $\varepsilon\beta$, $\varepsilon\theta$ and $\varepsilon\gamma$.

The corresponding distortion of the unperturbed ellipse (denoted $\mathbf{r}_0$) is

$$\mathbf{r}_o = \mathbf{r}_0 - \mathfrak{k}\varepsilon a\left(\frac{3z_\Lambda^3}{16(1-\Omega)^2(3-2\Omega)} + \frac{3(5-4\Omega)}{16z_\Lambda(1-\Omega)^2(1-2\Omega)} + \cdots\right)\exp(\mathrm{i}\Lambda) + \cdots \tag{6.49}$$

Due to the negative sign, both terms slightly compress the Moon's orbit along the Sun's direction, at the same time elongating it in the perpendicular direction (visualize this in the $\Lambda = 0$ frame; then rotate it back to Moon's Kepler frame). This is a somehow counter-intuitive result, opposite to the usual tidal effect.

Note that, when expressed in the *inertial* frame, the ε-proportional term reads

$$-\mathfrak{k}\varepsilon a\left(\frac{3z_\Lambda^3}{16(1-\Omega)^2(3-2\Omega)} + \frac{3(5-4\Omega)}{16z_\Lambda(1-\Omega)^2(1-2\Omega)} + \cdots\right)\exp(\mathrm{i}\Omega_0 t) \tag{6.50}$$

Similarly, for the corresponding Earth-Moon distance, we get

$$r = r_0 - \frac{3a\varepsilon(2-\Omega)\cos(2\omega - 2\Omega\varpi + 2\phi + 2\psi)}{2(1-\Omega)(3-2\Omega)(1-2\Omega)} + \cdots \simeq r_0 - a\varepsilon\cos(2\omega - 2\Omega\varpi + 2\phi + 2\psi) + \cdots \tag{6.51}$$

since the value of Ω is quite small.

6.2.5 Variation, evection and annual equation

Using the ε-accurate solution, one can make the corresponding adjustment to (1.75).

Firstly, the second line of (1.75) acquires, due to extending (1.74) by (6.50), a new term, equal to

$$-\varepsilon \text{Im}\left[\frac{3z_\Lambda^2}{16(1-\Omega)^2(3-2\Omega)}+\frac{3(5-4\Omega)}{16z_\Lambda^2(1-\Omega)^2(1-2\Omega)}\right]=\frac{3\varepsilon(7-10\Omega+4\Omega^2)\sin(2\omega-2\Omega_0 t+2\phi+2\psi)}{8(1-\Omega)^2(3-2\Omega)(1-2\Omega)}$$
$$\simeq\frac{7\varepsilon}{8}\sin(2\omega-2\Omega_0 t+2\phi+2\psi) \tag{6.52}$$

Secondly, we have to evaluate

$$t=2\int\sqrt{\frac{a}{\mu}}r\ \mathrm{d}s=\int\sqrt{\frac{a}{\mu}}\frac{r}{1-s_\text{p}'}\mathrm{d}\omega \tag{6.53}$$

to the ε^1 accuracy. Expanding the last integrand in ε, and simplifying each coefficient by keeping only the leading term of its Ω expansion, we get

$$\sqrt{\frac{a^3}{\mu}}(1+s_\text{p}')\left[1+\frac{\beta}{1+\beta^2}\left(z+\frac{1}{z}\right)-\varepsilon\cos(2\omega-2\Omega\varpi+2\phi+2\psi)\right]\simeq$$
$$\sqrt{\frac{a_\text{o}^3}{\mu}}\left[1+\frac{\beta}{1+\beta^2}\left(z+\frac{1}{z}\right)-\varepsilon\cos(2\omega-2\Omega\varpi+2\phi+2\psi)+\varepsilon+\frac{15\varepsilon}{4}\cos(2\Omega\varpi-2\phi-2\psi)+\frac{9\varepsilon\gamma}{4}\cos(\Omega\varpi)\right] \tag{6.54}$$

in view of (6.20) and

$$s_\text{p}'\simeq\varepsilon+\frac{15\varepsilon}{4}\cos(2\Omega\varpi-2\phi-2\psi) \tag{6.55}$$

Integrating (6.54) results in an updated version of (1.61), namely

$$t-t_0=\sqrt{\frac{a_\text{o}^3}{\mu}}(\omega+2\beta\sin\omega)+\varepsilon\sqrt{\frac{a_\text{o}^3}{\mu}}\left(\omega+\frac{15}{8\Omega}\sin(2\Omega\varpi-2\phi-2\psi)\right.$$
$$\left.-\frac{1}{2}\sin(2\omega-2\Omega\varpi+2\phi+2\psi)+\frac{9\gamma}{4\Omega}\sin(\Omega\varpi)\right) \tag{6.56}$$

where, recalling (6.22),

$$\beta\simeq\beta_\text{o}\left(1+\frac{15\varepsilon}{8\Omega}\cos(2\Omega_0 t-2\phi-2\psi)\right) \tag{6.57}$$

Defining

$$\tau\equiv\frac{t-t_0}{1-\varepsilon}\sqrt{\frac{\mu}{a_\text{o}^3}} \tag{6.58}$$

(note that Kepler's third law is no longer exact), and solving for ω, yields the following 'perturbed' version of (1.76):

$$\tau-2\beta_\text{o}\left(1+\frac{15\varepsilon}{8\Omega}\cos(2\Omega_0 t-2\phi_\text{sec}-2\psi_\text{sec})\right)\sin\left(\tau-2\beta_\text{o}-\frac{15\varepsilon}{8\Omega}\sin(2\Omega_0 t-2\phi_\text{sec}-2\psi_\text{sec})+\cdots\right)$$
$$-\frac{15\varepsilon}{8\Omega}\sin(2\Omega_0 t-2\phi_\text{sec}-2\psi_\text{sec})+\frac{\varepsilon}{2}\sin(2\tau-2\Omega_0 t+2\phi_\text{sec}+2\psi_\text{sec})-\frac{9\varepsilon\gamma}{4\Omega}\sin(\Omega_0 t)+\cdots \tag{6.59}$$

Substituting this into the second line of (1.75) results in

$$\begin{aligned}
&\tau - 2\beta_o \left(1 + \frac{15\varepsilon}{8\Omega}\cos(2\Omega_0 t - 2\phi_{\text{sec}} - 2\psi_{\text{sec}})\right)\sin\left(\tau - 2\beta_o - \frac{15\varepsilon}{8\Omega}\sin(2\Omega_0 t - 2\phi_{\text{sec}} - 2\psi_{\text{sec}}) + \cdots\right) \\
&-\frac{15\varepsilon}{8\Omega}\sin(2\Omega_0 t - 2\phi_{\text{sec}} - 2\psi_{\text{sec}}) + \frac{\varepsilon}{2}\sin(2\tau - 2\Omega_0 t + 2\phi_{\text{sec}} + 2\psi_{\text{sec}}) - \frac{9\varepsilon\gamma}{4\Omega}\sin(\Omega_0 t) \\
&+(\phi + \psi)_{\text{sec}} + (\phi + \psi)_{\text{osc}} \\
&-2\beta_o \left(1 + \frac{15\varepsilon}{8\Omega}\cos(2\Omega_0 t - 2\phi_{\text{sec}} - 2\psi_{\text{sec}})\right)\sin\left(\tau - 2\beta_o - \frac{15\varepsilon}{8\Omega}\sin(2\Omega_0 t - 2\phi_{\text{sec}} - 2\psi_{\text{sec}}) + \cdots\right) \\
&+\beta_o^2 \sin 2\tau - \tfrac{\theta_o^2}{4}\sin 2(\tau + \psi_{\text{sec}}) + \frac{7\varepsilon}{8}\sin(2\tau - 2\Omega_0 t + 2\phi_{\text{sec}} + 2\psi_{\text{sec}}) \qquad (6.60)
\end{aligned}$$

where we have also included correspondingly simplified (6.52). Expanding, simplifying, and re-arranging, one gets

$$\begin{aligned}
&\tau + (\phi + \psi)_{\text{sec}} - 4\beta_o \sin\tau + 5\beta_o^2 \sin 2\tau - \tfrac{\theta_o^2}{4}\sin 2(\tau + \psi_{\text{sec}}) + (\phi + \psi)_{\text{osc}} - \frac{15\varepsilon}{8\Omega}\sin(2\Omega_0 t - 2\phi_{\text{sec}} - 2\psi_{\text{sec}}) \\
&-\frac{15\varepsilon}{2\Omega}\beta_o \cos(2\Omega_0 t - 2\phi_{\text{sec}} - 2\psi_{\text{sec}})\sin\tau + \frac{15\varepsilon}{2\Omega}\beta_o \sin(2\Omega_0 t - 2\phi_{\text{sec}} - 2\psi_{\text{sec}})\cos\tau - \frac{9\varepsilon\gamma}{4\Omega}\sin(\Omega_0 t) \\
&+\frac{\varepsilon}{2}\sin(2\tau + 2\phi_{\text{sec}} + 2\psi_{\text{sec}} - 2\Omega_0 t) + \frac{7\varepsilon}{8}\sin(2\tau + 2\phi_{\text{sec}} + 2\psi_{\text{sec}} - 2\Omega_0 t) \qquad (6.61)
\end{aligned}$$

Note that $(\phi + \psi)_{\text{osc}}$ and the following term cancel out of (6.60).

The first two terms represent inertial-frame mean longitude; the next three terms are the 'unperturbed' effects of non-zero eccentricity and inclination, discussed in Chapter 1. The corresponding three coefficients (amplitudes), converted from radians to degrees, are -6.30°, 0.22° and -0.11° respectively (in good agreement with the usually quoted values of -6.29°, 0.21° and -0.11°).

The rest of (6.61) is due to Sun's perturbation. Combining the last two terms yields

$$\frac{11\varepsilon}{8}\sin(2\tau + 2\phi_{\text{sec}} + 2\psi_{\text{sec}} - 2\Omega_0 t) \qquad (6.62)$$

resulting in the so called VARIATION in Moon's longitude. Note that $(11/8)\varepsilon$ corresponds to 0.44° - not an impressive agreement with the observed value of 0.66°, but most of the error is due to assuming that $\Omega \simeq 0$ when computing the $\varepsilon/2$ and $(7/8)\varepsilon$ coefficients; using the original (6.51) - yet to be integrated - and (6.52) results in a lot more respectable 0.58°.The effect is driven by Moon's longitude *relative* to Sun's direction (it oscillates twice as fast).

The $\varepsilon\beta_o$-proportional terms of (6.61) further simplify to

$$-\frac{15\varepsilon}{2\Omega}\beta_o \sin(\tau - 2\Omega_0 t + 2\phi_{\text{sec}} + 2\psi_{\text{sec}}) \qquad (6.63)$$

The corresponding adjustment of Moon's longitude is called EVECTION. The formula has a relatively large amplitude of -1.32° (in reasonable agreement with the observed value of -1.27°) and oscillates with the difference between Moon's mean anomaly and $2\Omega_0 t - 2\phi_{\text{sec}} - 2\psi_{\text{sec}}$, the angle which controls eccentricity variation.

Finally, the second last term of (6.61), having an amplitude of -0.16° (in reasonable agreement with the observe value of -0.19°) is called ANNUAL EQUATION; it is driven by (and oscillates with) the Sun's distance.

6.2.6 Latitude

This time, all we have to do is to correspondingly modify (1.77), which now, in view of (6.46), reads

$$\theta_o \left(1 + \tfrac{3\varepsilon}{8\Omega(1-\Omega)}\cos(2\Omega_0 t - 2\phi)\right)\sin(\tau + \psi_{\text{sec}} + \psi_{\text{osc}} + \psi_0) - 4\beta_o\theta_o \cos(\tau + \psi_{\text{sec}} + \psi_0)\sin\tau \qquad (6.64)$$

where $\sin(\tau + \psi_{\text{sec}} + \psi_{\text{osc}} + \psi_0)$ expands to

$$\sin(\tau + \psi_{\text{sec}} + \psi_0) - \tfrac{3\varepsilon}{8\Omega(1-\Omega)} \sin(2\Omega_0 t - 2\phi)\cos(\tau + \psi_{\text{sec}} + \psi_0) \tag{6.65}$$

(6.64) can thus be simplified to read

$$\theta_o \sin(\tau + \psi_{\text{sec}} + \psi_0) - 4\beta_o\theta_o \cos(\tau + \psi_{\text{sec}} + \psi_0)\sin\tau + \tfrac{3\varepsilon}{8\Omega(1-\Omega)} \sin(\tau + \psi_{\text{sec}} + \psi_0 - 2\Omega_0 t + 2\phi) \tag{6.66}$$

where only the last term is due to Sun's perturbation. The three coefficients have the value of 5.13°, −0.56° and 0.16°, in good agreement with the observed values of 5.13°, −0.56° and 0.17°, respectively.

Note that $\tau + \psi_{\text{sec}} + \psi_0$ is Moon's mean longitude, and $\Omega_0 t - \phi$ is Sun's longitude, both *relative* to the nodal direction; τ is Moon's mean anomaly.

6.3 EXTENSIONS

To improve the accuracy of our solution, one has to include the contribution of

1. higher order terms in ε, β, γ and θ,
2. higher terms of the perturbing force (3.17),
3. Earth's oblateness,
4. Moon's (not perfectly spherical) shape
5. tidal effects
6. other planets (both directly on Moon's and indirectly on Earth's orbit)

Here, we briefly consider only two of these.

6.3.1 ε^2 terms

After one more iteration, the procedure returns the following results:

$$\begin{aligned}\frac{d\phi}{d\varpi} &= -\frac{3\varepsilon}{4} + \varepsilon^2\left(\frac{3}{32} + \frac{81}{16(1-2\Omega)} - \frac{9}{16(3-2\Omega)} - \frac{9(3-2\Omega)}{8(1-\Omega)^2}\right) \\ &+ \left[\frac{3\varepsilon}{4(1-\Omega)} - \varepsilon^2\left(\frac{189}{32(1-2\Omega)} - \frac{45}{32(3-2\Omega)} - \frac{3(41-24\Omega)}{32(1-\Omega)^2}\right)\right]\cos(2\Omega\varpi - 2\phi) + \cdots\end{aligned} \tag{6.67}$$

and

$$\begin{aligned}\frac{d(\phi+\psi)}{d\varpi} &= \frac{3\varepsilon}{4} + \varepsilon^2\left(\frac{15}{32} + \frac{9}{8(2-\Omega)} - \frac{9(392 - 1052\Omega + 940\Omega^2 - 279\Omega^3)}{128(1-\Omega)^4} + \frac{9(15-\Omega)}{16(3-2\Omega)^2}\right. \\ &\left. + \frac{81(1 - 14\Omega + 38\Omega^2 - 29\Omega^3)}{8(1-2\Omega)^4}\right) + \left[\frac{3\varepsilon}{4}\frac{5 - 5\Omega - 4\Omega^2}{(1-2\Omega)^2} - \varepsilon^2\left(\frac{3(140 - 201\Omega + 69\Omega^2)}{128(1-\Omega)^3}\right.\right. \\ &\left.\left. + \frac{9}{8(3-2\Omega)} - \frac{9(10 - 61\Omega + 66\Omega^2)}{32(1-2\Omega)^3}\right)\right]\cos(2\Omega\varpi - 2\phi - 2\psi) + \cdots\end{aligned} \tag{6.68}$$

where the three dots indicate 'dissonant' and fast-oscillating terms. Note that now the $' \to d../d\varpi$ conversion has been achieved by dividing the $d../ds$ derivatives by the ε-accurate version of $2(1 - s'_p)$.

With the help of (6.32), these differential equations yield the value of −18.705 and 8.723 years, respectively, for the length of the corresponding cycles. Extending the accuracy by yet another (ε^3-accurate) iteration improves the results only marginally to −18.704 and 8.726 years.

6.3.2 Effect of other planets

In this section, we consider only the *direct* effect of other planets on Moon's orbit. The resulting corrections are too small to improve our previous results; they are considered only out of theoretical interest.

For our purpose, it is sufficient to assume that each of the remaining planets has a circular orbit, coplanar with that of Earth. This implies, based on (3.20) and (4.12), that the corresponding time-averaged potential (over $\mathbf{r}_p$, in a heliocentric system) equals to

$$-\overline{\frac{m_p}{|\boldsymbol{\rho}-\mathbf{r}_p|}} = -\frac{m_p}{\rho}\sum_{n=0}^{\infty}\left(\frac{a_p}{\rho}\right)^n \frac{\int_0^{2\pi} P_n(\cos\gamma)\,\mathrm{d}\gamma}{2\pi} \tag{6.69}$$

for an inner planet, and

$$-\overline{\frac{m_p}{|\boldsymbol{\rho}-\mathbf{r}_p|}} \equiv -\frac{m_p}{a_p}\sum_{n=0}^{\infty}\left(\frac{\rho}{a_p}\right)^n \frac{\int_0^{2\pi} P_n(\cos\gamma)\,\mathrm{d}\gamma}{2\pi} \tag{6.70}$$

for an outer planet, where m_p and a_p is the planet's gravitational mass and semimajor axis, respectively, and $\rho \equiv |\boldsymbol{\rho}|$.

The negative gradient (with respect to $\boldsymbol{\rho}$) of each of these expressions yields the corresponding force, namely

$$-\frac{m_p\boldsymbol{\rho}}{\rho^3}\sum_{n=0}^{\infty}(n+1)\left(\frac{a_p}{\rho}\right)^n L_n \tag{6.71}$$

and

$$\frac{m_p\boldsymbol{\rho}}{\rho^3}\sum_{n=0}^{\infty} n\left(\frac{\rho}{a_p}\right)^{n+1} L_n \tag{6.72}$$

respectively, where $L_n \equiv \int_0^{2\pi} P_n(\cos\gamma)\,\mathrm{d}\gamma/2\pi$ (note that $L_n = 0$ when n is odd). To verify these, recall that

$$\nabla_{\boldsymbol{\rho}} f(\rho) = f'(\rho)\frac{\boldsymbol{\rho}}{\rho} \tag{6.73}$$

The heliocentric position of Moon is $\boldsymbol{\rho} = \mathbf{r} - \mathbf{R}$ ($\mathbf{r}$ and $\mathbf{R}$ being the geocentric positions of Moon and Sun, respectively). The corresponding distance ρ can be expanded, to a sufficient accuracy, as follows:

$$\rho \simeq R - \frac{\mathbf{r}\cdot\mathbf{R}}{R} \simeq A - \frac{\mathbf{r}\cdot\mathbf{R}}{A} \tag{6.74}$$

where the second term is small, compared to the first one. Substituting into (6.71) and (6.72) and time averaging over $\mathbf{R}$, we get the following final expression for the perturbing force

$$\begin{aligned}
&-\frac{m_p(\mathbf{r}-\mathbf{R})}{A^3}\sum_{n=0}^{\infty}(n+1)\left(1+(n+3)\frac{\mathbf{r}\cdot\mathbf{R}}{A^2}\right)\left(\frac{a_p}{A}\right)^n L_n \\
&\simeq \frac{m_p\mathbf{R}}{A^3}\sum_{n=0}^{\infty}(n+1)\left(\frac{a_p}{A}\right)^n L_n - \frac{m_p\mathbf{r}}{A^3}\sum_{n=0}^{\infty}(n+1)\left(\frac{a_p}{A}\right)^n L_n \\
&+\frac{m_p(\mathbf{r}\cdot\mathbf{R})\mathbf{R}}{A^5}\sum_{n=0}^{\infty}(n+1)(n+3)\left(\frac{a_p}{A}\right)^n L_n \\
&\simeq \frac{m_p\mathbf{r}}{2A^3}\sum_{n=0}^{\infty}(n+1)^2\left(\frac{a_p}{A}\right)^n L_n \equiv \frac{\varepsilon_p}{A^3}\mathbf{r}
\end{aligned} \tag{6.75}$$

and

$$\begin{aligned}
&\frac{m_p(\mathbf{r}-\mathbf{R})}{A^3}\sum_{n=0}^{\infty} n\left(1-(n-2)\frac{\mathbf{r}\cdot\mathbf{R}}{A^2}\right)\left(\frac{A}{a_p}\right)^{n+1} L_n \\
&\simeq -\frac{m_p\mathbf{R}}{A^3}\sum_{n=0}^{\infty} n\left(\frac{A}{a_p}\right)^{n+1} L_n + \frac{m_p\mathbf{r}}{A^3}\sum_{n=0}^{\infty} n\left(\frac{A}{a_p}\right)^{n+1} L_n \\
&\quad + \frac{m_p(\mathbf{r}\cdot\mathbf{R})\mathbf{R}}{A^5}\sum_{n=0}^{\infty} n(n-2)\left(\frac{A}{a_p}\right)^{n+1} L_n \\
&\simeq \frac{m_p\mathbf{r}}{2A^3}\sum_{n=0}^{\infty} n^2\left(\frac{A}{a_p}\right)^{n+1} L_n \equiv \frac{\varepsilon_p}{A^3}\mathbf{r}
\end{aligned} \tag{6.76}$$

for an inner and outer perturbing planet, respectively (m_p must now be expressed relative to Earth's mass).

This potential affects only the following two time derivatives

$$s_p' = \cdots + 2\frac{a^3}{A^3}\varepsilon_p \tag{6.77a}$$

$$\psi' = \cdots + 3\frac{a^3}{A^3}\varepsilon_p \tag{6.77b}$$

where the individual values of ε_p are:

Mercury	Venus	Mars	Jupiter	Saturn	rest
0.04	2.03	0.05	1.23	0.06	0.00

(6.78)

(their combined effect is of course additive; in the last two formulas, one should replace the individual ε_p's by their sum, equal to 3.41).

Due to the $a^3/A^3 \cong 1.68 \times 10^{-8}$ factor, this correction needs to be considered only in highly accurate solutions.

CHAPTER 7

Resonances

Abstract

Resonances occur when period of the perturbing force is commensurable with the perturbed body's orbital motion (2/1 resonance corresponds to a perturbing force whose one cycle is completed in two orbits of the perturbed body; two cycles are completed during one orbit in the case of 1/2 resonance, etc.). They constitute a rather special (and difficult) category of time-dependent perturbations, as we demonstrate using several examples of an asteroid perturbed by Jupiter's gravitational pull.

We start with the 1/1 resonance, which has some unique features, not shared by other resonances.

7.1 TROJAN ASTEROIDS (1/1 RESONANCE)

There are two distinct groups of asteroids, effectively sharing Jupiter's orbit. In this section we investigate the details of their interesting, dynamics [7], [14], [45].

Suppose there are two bodies (called *satellites*) orbiting the same primary, one of them is of gravitational mass M and the other is so small that its mass is negligible. According to the first line of (3.3), the equations describing their motion are

$$\ddot{\mathbf{r}} = -\mu\frac{\mathbf{r}}{r^3} - M\left(\frac{\mathbf{r}-\mathbf{R}}{|\mathbf{r}-\mathbf{R}|^3} + \frac{\mathbf{R}}{R^3}\right) \tag{7.1a}$$

$$\ddot{\mathbf{R}} = -(\mu + M)\frac{\mathbf{R}}{R^3} \tag{7.1b}$$

where μ is the gravitational mass of the primary, and $\mathbf{r}$ and $\mathbf{R}$ is the location of the massless and heavy satellite, respectively, with respect to the primary's center.

We would like to solve these equations assuming that the two satellites have nearly identical semimajor axes (a situation similar to an asteroid sharing an orbit with Jupiter; many such asteroids have been actually observed and are collectively referred to as Trojans, even though they consist of two distinct 'camps' which, individually, are called Trojans and Greeks).

To simplify the solution, we further assume that the two orbits are *coplanar*, and that the perturbing (heavy) satellite's orbit is *circular*, i.e.

$$\mathbf{R} = \mathfrak{k}\exp(\mathrm{i}t) \tag{7.2}$$

by our choice of units (i.e. its distance from the primary defines the unit of length, and its period equals 2π units of time). This makes $\mu + M$ equal to 1, implying $\mu = 1/(1+\varepsilon)$ and $M = \varepsilon/(1+\varepsilon)$, where $\varepsilon \equiv M/\mu$ (in our case $\varepsilon \simeq 0.001$, the ratio of Jupiter's mass to that of Sun).

Due to the planar assumption, all vectors of interest (including $\mathbf{r}$) are only two dimensional, having only the $\mathfrak{j}$ and $\mathfrak{k}$ components. This implies that the, originally quaternion, quantity $\mathbb{U}$ can now have only *complex* values (or, equivalently, $b = 0$, $\mathcal{S}(z) = 0$, $\theta = 0$ and $\phi = 0$, further implying that $\Gamma = 0$), thus greatly simplifying the corresponding algebra.

Furthermore, even vectors can now be represented by *complex* quantities, only *assumed* to be premultiplied by $\mathfrak{k}$. For example, (7.2) thus becomes (in this new notation)

$$\mathbf{R} = \exp(\mathrm{i}t) \tag{7.3}$$

Jan Vrbik

The massless satellite's location $\mathbf{r}$ (equal to $\overline{\mathbb{U}}\mathfrak{k}\mathbb{U}$) is now, with $\mathbb{U}$ complex, equal to $\mathbb{U}^2$ (since $\overline{\mathbb{U}}\mathfrak{k} = \mathfrak{k}\mathbb{U}$, and $\mathfrak{k}$ is dropped). Similarly, $\mathbf{rf}$ becomes $-\mathbf{r}^*\mathbf{f}$ (a complex quantity) etc. So, clearly, when dealing with a planar problem, we no longer need quaternions.

Our basic equation (1.29), in its *complex* form (which we emphasize by using $\mathcal{U}$ instead of $\mathbb{U}$), and with the perturbing force of (7.1a), thus reduces to

$$2\mathcal{U}'' - (2\mathcal{U}'\mathcal{U}'^* - 4a)\frac{\mathcal{U}}{r} - \mathcal{U}'\frac{a'}{a} + 4a\varepsilon\mathcal{U}\mathbf{r}^*\left(\frac{\mathbf{r}-\mathbf{R}}{|\mathbf{r}-\mathbf{R}|^3} + \frac{\mathbf{R}}{R^3}\right) = 0 \tag{7.4}$$

We can re-parametrize $\mathcal{U}$ by

$$\mathcal{U} = \sqrt{a}\exp[\mathrm{i}(s + \frac{\psi}{2})] \tag{7.5}$$

with a and ψ becoming the new (real) dependent variables (note that this implies $\mathbf{r} = a\exp[\mathrm{i}(2s+\psi)]$ and $r = a$); a and $2s + \psi$ are thus *polar* coordinates of the satellite's location. This deviates from our usual representation of $\mathcal{U}$, to which we return starting from Section 7.1.4.

Substituting (7.5) into (7.4), we obtain

$$\begin{aligned}&\left(\frac{a''}{a} - \frac{3}{2}(\frac{a'}{a})^2 + \mathrm{i}\psi'' + \mathrm{i}(1+\frac{\psi'}{2})\frac{a'}{a} - 4\psi' - (\psi')^2\right)\mathcal{U}\\ &= -4a\varepsilon\mathcal{U}\left(\frac{a^2 - a\exp[\mathrm{i}(t-2s-\psi)]}{(1+a^2-2a\cos[t-2s-\psi])^{3/2}} + a\exp[\mathrm{i}(t-2s-\psi)]\right)\end{aligned} \tag{7.6}$$

since

$$\mathcal{U}' = \left(\frac{1}{2}\frac{a'}{a} + \mathrm{i}(1+\frac{\psi'}{2})\right)\mathcal{U} \tag{7.7}$$

$$\mathcal{U}'' = \left(\frac{1}{2}\frac{a''}{a} - \frac{1}{4}(\frac{a'}{a})^2 + \mathrm{i}\frac{\psi''}{2} + \mathrm{i}(1+\frac{\psi'}{2})\frac{a'}{a} - (1+\frac{\psi'}{2})^2\right)\mathcal{U} \tag{7.8}$$

and

$$|\mathbf{r}-\mathbf{R}|^2 = r^2 + R^2 - \mathbf{r}^*\mathbf{R} - \mathbf{R}^*\mathbf{r} = 1 + a^2 - 2a\cos(t-2s-\psi) \tag{7.9}$$

Cancelling $\mathcal{U}$ throughout the equation, and separating the real and purely imaginary parts, results in

$$\frac{a''}{a} - \frac{3}{2}(\frac{a'}{a})^2 - 4\psi' - (\psi')^2 = -4\varepsilon\left(\frac{a^3 - a^2\cos\Psi}{(1+a^2-2a\cos\Psi)^{3/2}} + a^2\cos\Psi\right) \tag{7.10}$$

and

$$\psi'' + \frac{a'}{a} + \frac{1}{2}\frac{a'}{a}\psi' = -4\varepsilon\left(\frac{a^2\sin\Psi}{(1+a^2-2a\cos\Psi)^{3/2}} - a^2\sin\Psi\right) \tag{7.11}$$

where

$$\Psi \equiv 2s + \psi - t = \psi + 2s - 2\sqrt{1+\varepsilon}\int a^{3/2}\mathrm{d}s = \psi - 2\int(\sqrt{1+\varepsilon}a^{3/2} - 1)\mathrm{d}s \tag{7.12}$$

due to (1.28b), as our μ has the value of $1/(1+\varepsilon)$. Note that Ψ represents angular distance from the massless to the heavy satellite.

Introducing Ψ enables us to eliminate ψ as a dependent variable, replacing ψ' by

$$\Psi' + 2(\sqrt{1+\varepsilon}a^{3/2} - 1) \tag{7.13}$$

and ψ'' by

$$\Psi'' + 3\sqrt{1+\varepsilon}\sqrt{a}a' \tag{7.14}$$

In place of (7.10) and (7.11), we thus get

$$\frac{a''}{a} - \frac{3}{2}(\frac{a'}{a})^2 - 4\sqrt{1+\varepsilon}a^{3/2}\Psi' - (\Psi')^2 = -4\varepsilon\left(\frac{a^3 - a^2\cos\Psi}{(1+a^2-2a\cos\Psi)^{3/2}} + a^2\cos\Psi\right) + 4(1+\varepsilon)a^3 - 4 \tag{7.15}$$

and

$$\Psi'' + 4\sqrt{1+\varepsilon}\sqrt{a}a' + \frac{1}{2}\frac{a'}{a}\Psi' = -4\varepsilon\left(\frac{a^2\sin\Psi}{(1+a^2-2a\cos\Psi)^{3/2}} - a^2\sin\Psi\right)$$

7.1.1 Lagrange points

The fixed points of the last set of equations are:

1. $\Psi = 0$ and a solution to

$$(1+\varepsilon)a^3 - 1 = \varepsilon a^2\left(\frac{a-1}{|1-a|^3} + 1\right) \tag{7.16}$$

   or, equivalently, to

$$a = 1 + \sqrt[3]{\frac{\varepsilon a^2}{1+a+a^2+\varepsilon a^2}} \tag{7.17}$$

   when $a > 1$, and

$$a = 1 - \sqrt[3]{\frac{\varepsilon a^2}{1+a+a^2+\varepsilon a^2}} \tag{7.18}$$

   when $a < 1$. Each of the two equations yields exactly one solution (they can be easily obtained iteratively, starting with $a = 1$; in the case of $\varepsilon = 0.001$, we get 1.0709 and 0.9323 respectively). The corresponding locations are called Lagrange points L_1 and L_2.

2. $\Psi = \pi$ and a solution to

$$(1+\varepsilon)a^3 - 1 = \varepsilon a^2\left(\frac{a+1}{|1+a|^3} - 1\right) \tag{7.19}$$

   or, equivalently, to

$$a = \sqrt[3]{1 - \varepsilon a^2\left(1 + a - \frac{1}{(1+a)^2}\right)} \tag{7.20}$$

   This equation has a single solution, very close to 1 (in our case, it is 0.999417) corresponding to Lagrange point L_3.

3. Taking $\cos\Psi = a/2$ will also make the right hand side of (7.16) equal to zero, at the same time reducing the right hand side of (7.15) to $4(a^3 - 1)$. To make it equal to zero leads to $a = 1$, implying $\Psi = \pm\pi/3$ and thus defining the last two Lagrange points L_4 and L_5.

7.1.2 Stability analysis

In the vicinity of each fixed point, say (a_0, Ψ_0), the solution to (7.15) and (7.16) will have the form of $a = a_0 + c_1 e^{\lambda s}$ and $\Psi = \Psi_0 + c_2 e^{\lambda s}$, where c_1 and c_2 are small (so that we can discard higher powers of c_1 and c_2, keeping only linear terms), and λ is a constant.

For the Lagrange points L_4 and L_5, this yields

$$c_1\lambda^2 e^{\lambda s} - 4c_2\lambda\sqrt{1+\varepsilon}e^{\lambda s} = (12+3\varepsilon)c_1 e^{\lambda s} \pm 3\sqrt{3}\varepsilon c_2 e^{\lambda s} \tag{7.21a}$$

and

$$c_2\lambda^2 e^{\lambda s} + 4c_1\lambda\sqrt{1+\varepsilon}e^{\lambda s} = \pm 3\sqrt{3}\varepsilon c_1 e^{\lambda s} + 9\varepsilon c_2 e^{\lambda s} \tag{7.21b}$$

since

$$(12+3\varepsilon)(a-1) \pm 3\sqrt{3}\varepsilon(\Psi \mp \frac{\sqrt{3}}{2}) + \cdots \tag{7.21c}$$

and

$$\pm 3\sqrt{3}\varepsilon(a-1) + 9\varepsilon(\Psi \mp \frac{\sqrt{3}}{2}) + \cdots \tag{7.21d}$$

are the generalized Taylor expansion of the right hand sides of (7.15) and (7.16), respectively, at $a = 1$ and $\Psi = \pm\sqrt{3}/2$.

The two equations (7.21a) and (7.21b) can be re-written in the following matrix form

$$\begin{bmatrix} \lambda^2 - 12 - 3\varepsilon & -4\lambda\sqrt{1+\varepsilon} \mp 3\sqrt{3}\varepsilon \\ 4\lambda\sqrt{1+\varepsilon} \mp 3\sqrt{3}\varepsilon & \lambda^2 - 9\varepsilon \end{bmatrix} \mathbf{c} = \mathbf{0} \tag{7.22}$$

where

$$\mathbf{c} \equiv \begin{bmatrix} c_1 \\ c_2 \end{bmatrix} \tag{7.23}$$

Obviously, there is a non-zero solution (in terms of c_1 and c_2) only when the coefficient matrix is singular, i.e. when its determinant is equal to zero, implying:

$$\lambda^4 + 4(1+\varepsilon)\lambda^2 + 108\varepsilon = 0 \tag{7.24}$$

The roots of this characteristic polynomial are the four solutions to

$$\lambda^2 = -2(1+\varepsilon) \pm 2\sqrt{(1+\varepsilon)^2 - 27\varepsilon} \tag{7.25}$$

All four are purely imaginary when $(1+\varepsilon)^2 > 27\varepsilon$ or, equivalently, when $\varepsilon < 25/2 - \sqrt{(25/2)^2 - 1} \simeq 0.04006$. This implies that the corresponding fixed points are, for these values of ε, stable centers. On the other hand, when $\varepsilon > 0.04006$, two of the four complex roots have a positive real part, implying instability of the corresponding solution.

By a similar analysis one can show that the other three Lagrange points (L_1, L_2 and L_3) are unstable for any value of ε. This is why we now concentrate on exploring the stable solutions at L_4 and L_5.

The actual motion is, in this linearized approximation, a superposition of two elliptical oscillation:

1. one 'slow', with $\lambda \simeq \pm\sqrt{27\varepsilon}\mathrm{i}$ and

$$\mathbf{c} \propto \begin{bmatrix} \sqrt{3\varepsilon}\cos[\sqrt{27\varepsilon}(s - s_p)] \\ -\sin[\sqrt{27\varepsilon}(s - s_p)] \end{bmatrix} \tag{7.26}$$

representing (when converted to rectangular coordinates) a small 'tadpole' orbit,

2. the other 'fast', with $\lambda \simeq \pm 2\mathrm{i}$ and

$$\mathbf{c} \propto \begin{bmatrix} \cos[2(s - s_p)] \\ -2\sin[2(s - s_p)] \end{bmatrix} \tag{7.27}$$

corresponding to variations in a due to nonzero eccentricity (recall that now a represents distance from the primary, not the semimajor axis).

7.1.3 Zero-eccentricity solution

In this segment, we find *global* solutions (as opposed to solutions near an equilibrium point) to (7.10) and (7.11). To make the task easier, we assume (in addition to our earlier, circular and planar assumptions), that the perturbed body (asteroid) has *zero eccentricity.* This time, zero eccentricity is *not* synonymous with 'circular'; instead, we get several distinct classes of possible solutions, their shapes being described as 'horseshoe', 'hourglass', and 'tadpole'. Let us have a closer look at these.

The new assumption implies that both a and ψ of are lacking the 'fast'-oscillating component, and are therefore slow-changing quantities. This further implies that both a' and ψ' must be 'small' (proportional to $\sqrt{\varepsilon}$ when close to an equilibrium point, as shown in the previous segment), and their higher derivatives (a'', ψ'', etc.) are smaller yet (of a correspondingly higher order of ε).

The two equations can now be re-written ('solved' for ψ' and a'/a) as follows:

$$\psi' = \varepsilon\left(\frac{a^3 - a^2\cos\Psi}{(1+a^2-2a\cos\Psi)^{3/2}} + a^2\cos\Psi\right) + \frac{1}{4}\left(\frac{a'}{a}\right)' - \frac{1}{8}\left(\frac{a'}{a}\right)^2 - \frac{1}{4}(\psi')^2 \tag{7.28}$$

and

$$\frac{a'}{a} = -4\varepsilon\left(\frac{a^2\sin\Psi}{(1+a^2-2a\cos\Psi)^{3/2}} - a^2\sin\Psi\right) - \psi'' - \frac{1}{2}\frac{a'}{a}\psi' \tag{7.29}$$

Starting with the initial values of $a'/a = 0$ and $\psi' = 0$, the two equations can be solved, *iteratively*, for a'/a and ψ', to any order of accuracy. One must remember that

$$\Psi' = \psi' - 2(\sqrt{1+\varepsilon}a^{3/2} - 1) \tag{7.30}$$

which is based on (7.13). Note that the last term is proportional to $\varepsilon^{1/2}$ for orbits centered on one of the two stable Lagrange points (TADPOLE orbits), and proportional to $\varepsilon^{1/3}$ for orbits stretching between both stable points (HORSESHOE orbits), as demonstrated shortly.

In the end, the resulting ψ' can be easily converted to the more convenient Ψ'.

For our purpose, it is adequate to use only the first two nonzero terms of the a' and Ψ' expansions, namely:

$$\Psi' = -2(\sqrt{1+\varepsilon}a^{3/2} - 1) + \varepsilon\left(\frac{a^3 - a^2\cos\Psi}{(1+a^2-2a\cos\Psi)^{3/2}} + a^2\cos\Psi\right) + \cdots \tag{7.31}$$

which is (7.30) with only the first term from the right hand side of (7.28), and

$$\begin{aligned}\frac{a'}{a} = &-4\varepsilon\left(\frac{a^2\sin\Psi}{(1+a^2-2a\cos\Psi)^{3/2}} - a^2\sin\Psi\right) \\ &+2(\sqrt{1+\varepsilon}a^{3/2} - 1)\varepsilon\frac{\partial}{\partial\Psi}\left(\frac{a^3 - a^2\cos\Psi}{(1+a^2-2a\cos\Psi)^{3/2}} + a^2\cos\Psi\right) + \cdots\end{aligned} \tag{7.32}$$

obtained from the first two terms of the right hand side of (7.29), accordingly simplified.

To the same accuracy, the following expression is a constant of motion

$$K = 4\sqrt{1+\varepsilon}a^{1/2} + \frac{2}{a} + 4\varepsilon\left(\frac{1}{(1+a^2-2a\cos\Psi)^{1/2}} - a\cos\Psi\right)$$
$$+2(\sqrt{1+\varepsilon}a^{3/2}-1)\varepsilon\left(\frac{a^2-a\cos\Psi}{(1+a^2-2a\cos\Psi)^{3/2}} + a\cos\Psi\right) + \cdots \tag{7.33}$$

since

$$a\frac{\partial K}{\partial \Psi} = \frac{a'}{a} \tag{7.34a}$$
$$a^2\frac{\partial K}{\partial a} = -\Psi' \tag{7.34b}$$

(as can be verified easily), implying

$$a\frac{\partial K}{\partial \Psi}\Psi' + a\frac{\partial K}{\partial a}a' = \frac{a'}{a}\Psi' - \Psi'\frac{a'}{a} = 0 \tag{7.35}$$

The contours of K yield the corresponding (approximate) orbits. They are displayed in Figure 7.1 for $\varepsilon = 0.01$ (an order of magnitude bigger than the actual value, to have the individual types of orbit stand out). Recall that a and Ψ represent the orbit's polar coordinates in a rotating frame of the heavy satellite (which is located at $a = 1$ and $\Psi = 0$); in the plot, they have been converted to rectangular coordinates.

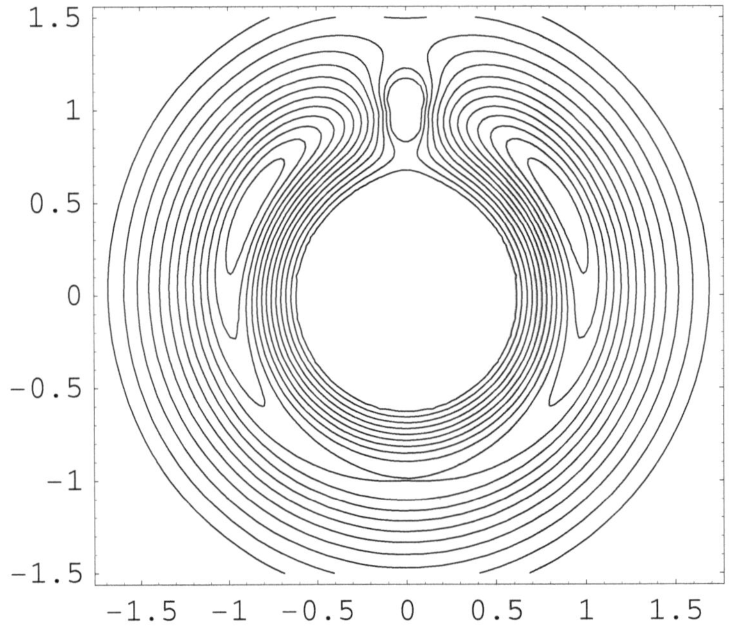

Figure 7.1 Tadpole and horseshoe orbits (restricted problem).

Note the formation of *tadpole* and *horseshoe* orbits (when close to the separatrix, these are also called *hourglass* orbits), having the maximum 'thickness' (in terms of a) of about $2\varepsilon^{1/2}$ and $2\varepsilon^{1/3}$ respectively.

Our graph also correctly indicates the possibility of inner and outer *passing* orbits, even though their description near $\Psi = 0$ is only qualitative (our iterative procedure does not converge in that region). Finally, we can also see some highly distorted (for the same reason) '*captured*' orbits, where the massless satellite simply orbits the heavy one.

7.1.4 Non-zero eccentricity

To build a solution with nonzero eccentricity, we have to use (1.105) with $b = 0$, $\mathcal{S}(z) = 0$, $\theta = 0$ and $\phi = 0$, namely

$$\mathcal{U} = \sqrt{\frac{a}{1+\beta^2}}\left[q + \beta q^{-1} + q\mathcal{D}(z)\right]\exp(\mathrm{i}\frac{\psi}{2}) \tag{7.36}$$

where, unlike in the previous segment, a and ψ revert back to being the usual orbital parameters.

We now construct the *first order* (in ε) solution, assuming that the resulting eccentricity is small, and expanding all quantities accordingly (up to terms linear in β); we also replace a by 1 in all β-proportional terms.

Substituting the zero-order solution into (7.4), expressing the answer in Kepler frame, and multiplying it by

$$-\frac{\mathcal{U}_0^*}{2(1+\beta z)} \tag{7.37}$$

yields

$$\mathcal{Q}(z) = -\frac{2a\varepsilon}{1+\beta z}\left(\frac{r_o^2 - \mathbf{r}_o^*\mathbf{R}_o}{|\mathbf{r}_o - \mathbf{R}_o|^3} + \frac{\mathbf{r}_o^*\mathbf{R}_o}{R_o^3}\right) \tag{7.38}$$

where

$$\mathbf{R}_o = \exp(\mathrm{i}t)\exp(-\mathrm{i}\psi) \simeq z\exp(-\mathrm{i}\Psi)\left(1 + \beta z - \frac{\beta}{z} + \cdots\right) \tag{7.39a}$$

$$\mathbf{r}_o = a(z + 2\beta) + \cdots \tag{7.39b}$$

$$r_o^2 = a^2\left(1 + 2\beta z + \frac{2\beta}{z}\right) + \cdots \tag{7.39c}$$

since

$$\begin{aligned} t &= 2\sqrt{1+\varepsilon}\int a^{3/2}\left(1 + \frac{2\beta}{1+\beta^2}\cos[2(s - s_p)]\right)\mathrm{d}s \\ &\simeq 2(s - s_p) + 2\int (a^{3/2} - 1)\mathrm{d}s + 2s_p + 2\beta\sin[2(s - s_p)] + \cdots \end{aligned} \tag{7.39d}$$

and, by definition,

$$\Psi \equiv \psi - 2s_p - 2\int (a^{3/2} - 1)\mathrm{d}s \tag{7.39e}$$

The last quantity still represents the asteroid's angular distance from Jupiter.

(7.39a) and (7.39b) further imply that

$$\mathbf{r}_o^*\mathbf{R}_o = a\left(1 + 3\beta z - \frac{\beta}{z}\right)\exp(-\mathrm{i}\Psi) + \cdots \tag{7.40}$$

and

$$\begin{aligned} |\mathbf{r}_o - \mathbf{R}_o|^{-3} &= \left[1 + a^2 - 2a\cos\Psi + \beta\left(2z + \frac{2}{z} - (3z - \frac{1}{z})\exp(-\mathrm{i}\Psi) + (z - \frac{3}{z})\exp(\mathrm{i}\Psi)\right) + \cdots\right]^{-3/2} \\ &\simeq \frac{1}{(1 + a^2 - 2a\cos\Psi)^{3/2}} - \frac{\frac{3}{2}\beta\left(2z + \frac{2}{z} - (3z - \frac{1}{z})\exp(-\mathrm{i}\Psi) + (z - \frac{3}{z})\exp(\mathrm{i}\Psi)\right)}{(2 - 2\cos\Psi)^{5/2}} + \cdots \end{aligned} \tag{7.41}$$

Finally, from (7.38)

$$\begin{aligned}\mathcal{Q}(z) \simeq & -2a\varepsilon\left(\frac{a^2 - a\exp(-\mathrm{i}\Psi)}{(1+a^2-2a\cos\Psi)^{3/2}} + a\exp(-\mathrm{i}\Psi)\right) \\ & -2\varepsilon\beta\left(\frac{z+\frac{2}{z}-(2z-\frac{1}{z})\exp(-\mathrm{i}\Psi)}{(2-2\cos\Psi)^{3/2}} + (2z-\frac{1}{z})\exp(-\mathrm{i}\Psi)\right) \\ & +3\varepsilon\beta\frac{(1-\exp(-\mathrm{i}\Psi))\left(2z+\frac{2}{z}-(3z-\frac{1}{z})\exp(-\mathrm{i}\Psi)+(z-\frac{3}{z})\exp(\mathrm{i}\Psi)\right)}{(2-2\cos\Psi)^{5/2}} + \cdots\end{aligned} \tag{7.42}$$

The last expression implies that

$$\frac{a'}{a} \simeq 2\mathrm{Im}(\mathcal{Q}_0 - \beta\mathcal{Q}_{-1}) = -4\varepsilon a^2\sin\Psi\left(\frac{1}{(1+a^2-2a\cos\Psi)^{3/2}} - 1\right) + \cdots \tag{7.43a}$$

which agrees with the first term of (7.32),

$$\frac{\beta'}{\beta} \simeq -\frac{\mathrm{Im}(\mathcal{Q}_1 + 3\beta\mathcal{Q}_0 + 3\mathcal{Q}_{-1} + \beta\mathcal{Q}_{-2})}{4\beta} = \varepsilon\sin\Psi\left(\frac{1}{(2-2\cos\Psi)^{3/2}} - 1\right) + \cdots \simeq -\frac{1}{4}\frac{a'}{a} \tag{7.43b}$$

(note that the β oscillations mimic those of a - having the opposite phase and smaller amplitude), and

$$\psi' \simeq \frac{\mathrm{Re}(-\mathcal{Q}_1 + \beta\mathcal{Q}_0 + 3\mathcal{Q}_{-1} + \beta\mathcal{Q}_{-2})}{4\beta} = \frac{\varepsilon}{2}\left(\frac{9+5\cos\Psi}{(2-2\cos\Psi)^{3/2}} + 4\cos\Psi\right) + \cdots \tag{7.43c}$$

At the stable Lagrange points, the last equation corresponds to a uniform circulation of the pericenter at the rate of $(27/4)\varepsilon$.

The final conclusion is that, in terms of a and Ψ, the solution remains practically the same as in the $\beta = 0$ case; the only visible effect of a non-zero β is to introduce periodic variation in r, as shown in Figure 7.2 (again, using the rotating frame, rectangular coordinates, and $\varepsilon = 0.01$; demonstrating the new shape of a tadpole orbit).

7.1.5 Extensions

When the heavy satellite has a small non-zero eccentricity γ, we can easily extend our results by terms linear in γ (in these, we also replace a by 1).

One can readily verify that $\mathbf{R}_o$ needs to be extended, in analogy with (6.3), by

$$\cdots + \gamma\frac{z^2\exp(-2\mathrm{i}\Psi + \mathrm{i}\psi) - 3\exp(-\mathrm{i}\psi)}{2} \tag{7.44}$$

$\mathbf{r}_o^*\mathbf{R}_o$ by

$$\cdots + \gamma\left(\frac{z}{2}\exp(-2\mathrm{i}\Psi + \mathrm{i}\psi) - \frac{3}{2z}\exp(-\mathrm{i}\psi)\right) \tag{7.45}$$

R_o^2 by

$$\cdots - \gamma\left(z\exp(-\mathrm{i}\Psi + \mathrm{i}\psi) + \frac{1}{z}\exp(\mathrm{i}\Psi - \mathrm{i}\psi)\right) \tag{7.46}$$

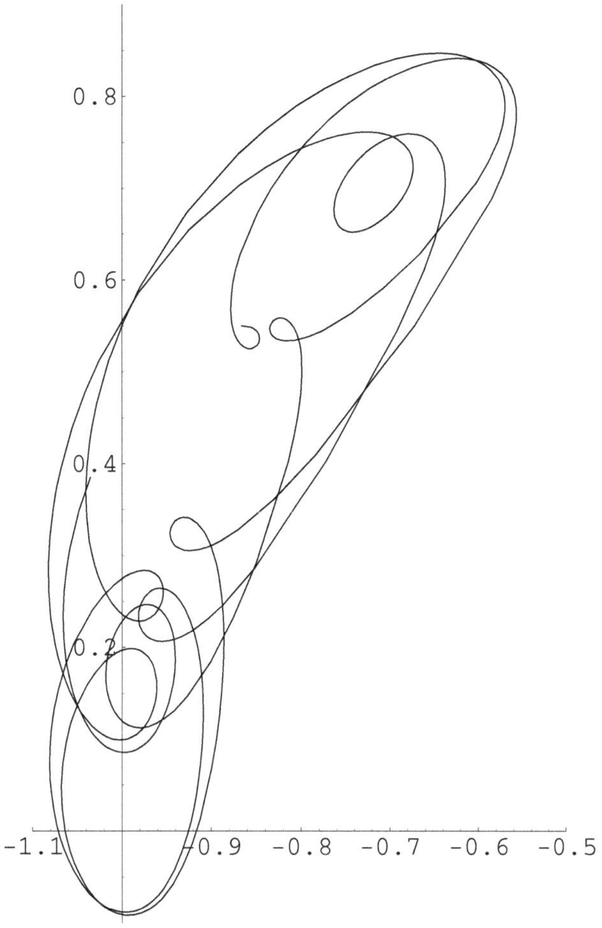

Figure 7.2 Tadpole orbit with non-zero eccentricity.

and $|\mathbf{r}_o - \mathbf{R}_o|^{-3}$ by

$$\cdots + \frac{3}{2}\gamma z \exp(i\psi) \frac{\left(\exp(-i\Psi) + \frac{1}{2}\exp(-2i\Psi) - \frac{3}{2}\right)}{(2 - 2\cos\Psi)^{5/2}} + \frac{3}{2}\gamma \frac{1}{z} \exp(-i\psi) \frac{\left(\exp(i\Psi) + \frac{1}{2}\exp(2i\Psi) - \frac{3}{2}\right)}{(2 - 2\cos\Psi)^{5/2}} \tag{7.47}$$

This results in $Q(z)$ (see Eq. 7.38) acquiring an extra

$$\begin{aligned}
&\cdots - \varepsilon\gamma \left(z\exp(-2i\Psi + i\psi) - \frac{3}{z}\exp(-i\psi) \right) \left(1 - \frac{1}{(2 - 2\cos\Psi)^{3/2}} \right) \\
&-3\varepsilon\gamma \exp(-i\Psi) \left(z\exp(-i\Psi + i\psi) + \frac{1}{z}\exp(i\Psi - i\psi) \right) \\
&-3\varepsilon\gamma z \frac{(1 - \exp(-i\Psi))\exp(i\psi)\left(\exp(-i\Psi) + \frac{1}{2}\exp(-2i\Psi) - \frac{3}{2}\right)}{(2 - 2\cos\Psi)^{5/2}} \\
&-3\varepsilon\gamma \frac{1}{z} \frac{(1 - \exp(-i\Psi))\exp(-i\psi)\left(\exp(i\Psi) + \frac{1}{2}\exp(2i\Psi) - \frac{3}{2}\right)}{(2 - 2\cos\Psi)^{5/2}} \\
&= \cdots - 4\varepsilon\gamma z \exp(-2i\Psi + i\psi) - \frac{1}{2}\varepsilon\gamma z \exp(i\psi) \frac{9\exp(-i\Psi) + \exp(-2i\Psi)}{(2 - 2\cos\Psi)^{3/2}} \\
&+ \frac{3}{2}\varepsilon\gamma \frac{1}{z} \exp(-i\psi) \frac{1 + \exp(i\Psi)}{(2 - 2\cos\Psi)^{3/2}}
\end{aligned} \tag{7.48}$$

implying no change in a'/a, an additional term

$$\cdots - \varepsilon\frac{\gamma}{\beta}\sin(2\Psi - \psi) - \frac{\varepsilon}{8}\frac{\gamma}{\beta} \frac{\sin(2\Psi - \psi) + 18\sin(\Psi - \psi) - 9\sin\psi}{(2 - 2\cos\Psi)^{3/2}} \tag{7.49}$$

in β'/β, and a similar extra term

$$\cdots + \varepsilon\frac{\gamma}{\beta}\cos(2\Psi - \psi) + \frac{\varepsilon}{8}\frac{\gamma}{\beta} \frac{\cos(2\Psi - \psi) + 18\cos(\Psi - \psi) + 9\cos\psi}{(2 - 2\cos\Psi)^{3/2}} \tag{7.50}$$

in ψ'.

Note that the last two equations yield, at the stable Lagrange points ($\Psi = \pm\pi/3$)

$$\beta' = \varepsilon\gamma \left(\frac{27}{16}\sin\psi \mp \frac{27\sqrt{3}}{16}\cos\psi \right) \tag{7.51}$$

and

$$\psi' = \frac{27}{4}\varepsilon + \varepsilon\frac{\gamma}{\beta} \left(\frac{27}{16}\cos\psi \pm \frac{27\sqrt{3}}{16}\sin\psi \right) \tag{7.52}$$

The family of solutions to the last two equations are the contours of

$$\frac{27}{8}\varepsilon\beta^2 + \varepsilon\gamma\beta \left(\frac{27}{16}\cos\psi \pm \frac{27\sqrt{3}}{16}\sin\psi \right) \tag{7.53}$$

displayed (for L_4) in Figure 7.3, and produced with $\varepsilon = 0.001$ and $\gamma = 0.05$ (roughly Jupiter's actual values). Note the stable center at $\psi = 4/3 \cdot \pi$ ($2/3 \cdot \pi$ in case of L_5) and $\beta = \gamma/2$.

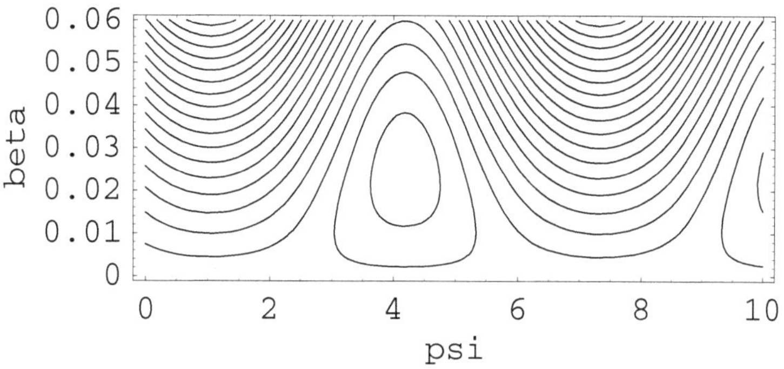

Figure 7.3 Solutions near L_4 (Jupiter's eccentricity = 0.05).

Secondly, we remove the assumption of coplanar orbits, introducing a (small) inclination θ of the two orbital planes. This means we have to go back to the quaternionic version of the basic equation (1.29).

Evaluating all quantities only up to the linear terms in θ (discarding terms proportional to β and γ - those have been dealt with before) we obtain:

$$\mathbf{R}_o = \mathfrak{k} z \exp(-\mathrm{i}\Psi) - \frac{\theta}{2}\left(z\exp(-\mathrm{i}\Psi + \mathrm{i}\psi) - \frac{1}{z}\exp(\mathrm{i}\Psi - \mathrm{i}\psi)\right) + \cdots \tag{7.54}$$

where

$$\Psi \equiv \psi + \phi - 2s_p - 2\int (a^{3/2} - 1)\mathrm{d}s \tag{7.55}$$

(note that, in this approximation, $\psi + \phi$ is the apocenter's direction - up to now, this used to be called ψ).

Together with

$$\overline{\mathbf{r}_o}\mathbf{R}_o = a\exp(-\mathrm{i}\Psi) + \mathfrak{k}\frac{\theta}{2}\left(z^2\exp(-\mathrm{i}\Psi + \mathrm{i}\psi) - \exp(\mathrm{i}\Psi - \mathrm{i}\psi)\right) \tag{7.56}$$

and

$$\overline{\mathbf{R}_o}\mathbf{r}_o = a\exp(-\mathrm{i}\Psi) - \mathfrak{k}\frac{\theta}{2}\left(z^2\exp(-\mathrm{i}\Psi + \mathrm{i}\psi) - \exp(\mathrm{i}\Psi - \mathrm{i}\psi)\right) \tag{7.57}$$

(7.54) implies that R_o^2 and $|\mathbf{r}_o - \mathbf{R}_o|^{-3}$ remain unchanged.

As a result, we get no θ-proportional contribution to $\mathcal{Q}(z)$. This implies that the dynamics of a, β and $\psi + \phi$ (the old ψ) are, within this accuracy, not affected by a small inclination of the two orbits.

The only nonzero terms are the consequence of expanding $\mathcal{W}(z)$, as follows

$$4a\varepsilon r_o\left(\frac{\mathbf{r}_o - \mathbf{R}_o}{|\mathbf{r}_o - \mathbf{R}_o|^3} + \frac{\mathbf{R}_o}{R_o^3}\right) \simeq 2\varepsilon\theta\left(z\exp(-\mathrm{i}\Psi + \mathrm{i}\psi) - \frac{1}{z}\exp(\mathrm{i}\Psi - \mathrm{i}\psi)\right)\left(\frac{1}{(2 - 2\cos\Psi)^{3/2}} - 1\right) + \cdots \tag{7.58}$$

From this, we obtain

$$Z_1 \simeq -\frac{1}{2}\mathrm{Im}\mathcal{W}_1 = \varepsilon\theta\sin(\Psi - \psi)\left(\frac{1}{(2 - 2\cos\Psi)^{3/2}} - 1\right) + \cdots \tag{7.59a}$$

$$Z_2 \simeq -\frac{1}{2}\mathrm{Re}\mathcal{W}_1 = -\varepsilon\theta\cos(\Psi - \psi)\left(\frac{1}{(2 - 2\cos\Psi)^{3/2}} - 1\right) + \cdots \tag{7.59b}$$

implying

$$\theta' = Z_1 \cos\psi - Z_2 \sin\psi \simeq -\varepsilon\theta \sin\Psi \left(1 - \frac{1}{(2-2\cos\Psi)^{3/2}}\right) + \cdots \tag{7.60a}$$

and

$$\phi' = \frac{Z_1 \sin\psi + Z_2 \cos\psi}{\sin\theta} \simeq \varepsilon \cos\Psi \left(1 - \frac{1}{(2-2\cos\Psi)^{3/2}}\right) + \cdots \tag{7.60b}$$

Note that (similarly to β'/β)

$$\frac{\theta'}{\theta} \simeq -\frac{1}{4}\frac{a'}{a} \tag{7.61}$$

and that both θ' and ϕ' equal to zero at the two stable Lagrange points.

7.2 HECUBA GAP (2/1 RESONANCE)

Similarly to the 1/1 resonance, the best and most numerous examples of other resonances appear in the asteroid belt. As we already know, the predominant perturbing body is Jupiter, which causes a noticeable decrease in the number of asteroids (the so called KIRKWOOD GAPS) in and near several resonances (2/1, 3/1, 5/2 and 7/3 being the most pronounced).

In this section, we discuss the details of the 2/1 resonance, which clears the so called HECUBA GAP, in the region where the orbital period of the massless satellite is *double* that of the perturbing body.[1] Later, our results are extended to cover resonances of other types.

We choose our units so that the satellite's a in exact resonance (we will take this to be also its the average value a_o), and μ, are both equal to 1 (this is slightly different from the last section, but more convenient for the current situation).

7.2.1 Circular perturbing orbit

To explain the dynamics of the resulting motion, we again start by assuming that the perturbing body has a circular orbit, and that the two orbits are *coplanar.*

We can thus use the complex equations of Section 7.1.4, which hold almost unchanged, except for:

1. the orbit of the heavy satellite is now described by

$$\mathbf{R} = 2^{2/3}(1+\varepsilon)^{1/3} \exp\left(\mathrm{i}\frac{t}{2}\right) \equiv A \exp\left(\mathrm{i}\frac{t}{2}\right) \tag{7.62}$$

2. which, converted to Kepler's frame of the *perturbed* satellite, reads

$$\mathbf{R}_o \simeq Aq \exp(-\mathrm{i}\Psi) \left(1 + \frac{\beta}{2}z - \frac{\beta}{2z} + \cdots\right) \tag{7.63}$$

where $q \equiv \exp[\mathrm{i}(s - s_p)]$ and

$$\Psi \equiv 2\psi - \chi \tag{7.64}$$

with χ as in (2.44) - slightly simplified by our $a_o = \mu = 1$ choice. Note that Ψ is an analog of the RESONANCE VARIABLE of Section 7.1.4 (in general, for a k/ℓ resonance, this variable equals to ψ multiplied by k, minus χ multiplied by ℓ). To a good approximation, Ψ is proportional to the perturbed body's longitude of apocenter at conjunction.

[1]Further reading: [17], [25], [35] and [43].

3.

$$\mathbf{r}_o^*\mathbf{R}_o \simeq A\left(a+\frac{5}{2}\beta z-\frac{\beta}{2z}\right)\frac{\exp(-\mathrm{i}\frac{\Psi}{2})}{q} \equiv$$
$$A\left(a+\frac{5}{2}\beta z-\frac{\beta}{2z}\right)\exp(-\mathrm{i}\xi) \tag{7.65}$$

where q was replaced by $\exp(\mathrm{i}\xi-\mathrm{i}\frac{\Psi}{2})$, which also means that $z=\exp(2\mathrm{i}\xi-\mathrm{i}\Psi)$. This implies

$$|\mathbf{r}_o-\mathbf{R}_o|^{-3} \simeq \frac{1}{(A^2+a^2-2Aa\cos\xi)^{3/2}} - \frac{\frac{3}{2}\beta\left(2z+\frac{2}{z}+A\exp(\mathrm{i}\xi)(\frac{z}{2}-\frac{5}{2z})+A\exp(-\mathrm{i}\xi)(\frac{1}{2z}-\frac{5z}{2})\right)}{(A^2+1-2A\cos\xi)^{5/2}} \tag{7.66}$$

and,

4. based on (7.38)

$$\begin{aligned}\mathcal{Q}(z) \simeq & -2\varepsilon\frac{a^2}{A^2}\exp(-\mathrm{i}\xi)-2\varepsilon a\frac{a^2-Aa\exp(-\mathrm{i}\xi)}{(A^2+a^2-2Aa\cos\xi)^{3/2}}\\ & -\varepsilon\frac{\beta}{A^2}\exp(-\mathrm{i}\xi)(3z-\frac{1}{z})-\varepsilon\beta\frac{2z+\frac{4}{z}-A\exp(-\mathrm{i}\xi)(3z-\frac{1}{z})}{(A^2+1-2A\cos\xi)^{3/2}}\\ & +3\varepsilon\beta\left(1-A\exp(-\mathrm{i}\xi)\right)\frac{\left(2z+\frac{2}{z}+A\exp(\mathrm{i}\xi)(\frac{z}{2}-\frac{5}{2z})+A\exp(-\mathrm{i}\xi)(\frac{1}{2z}-\frac{5z}{2})\right)}{(A^2+1-2A\cos\xi)^{5/2}}\end{aligned} \tag{7.67}$$

Assuming the orbital elements fixed (as we always do at this point), we note that $\mathcal{Q}(z)$ is a function of not only $z=q^2$, but also of q (via ξ). This means that its expansion may now be expressed (in terms of powers of q) as

$$\mathcal{Q}(z)=\cdots+\frac{\mathcal{Q}_{-3/2}}{q^3}+\frac{\mathcal{Q}_{-1}}{q^2}+\frac{\mathcal{Q}_{-1/2}}{q}+\mathcal{Q}_0+q\mathcal{Q}_{1/2}+q^2\mathcal{Q}_1+q^3\mathcal{Q}_{3/2}+q^4\mathcal{Q}_2+\cdots \tag{7.68}$$

with the individual coefficients computed via the obvious modification of (4.21).

Using the averaging principle (Section 2.4), we ignore the contribution of half-integer terms. Note that these would need to be included to extend the technique beyond the first-order-in-ε accuracy (something we do not intent to do here).

We thus compute only

$$\begin{aligned}&\mathcal{Q}_{-1}=\\ &\oint_{C_0} q^2\mathcal{Q}(q)\frac{\mathrm{d}q}{2\pi\mathrm{i}q} \equiv \frac{1}{2\pi}\int_0^{2\pi}\mathcal{Q}(\xi)\exp(2\mathrm{i}\xi-\mathrm{i}\Psi)\mathrm{d}\xi \simeq\\ &-2\varepsilon a^2\exp(-\mathrm{i}\Psi)\left(a\mathsf{F}(\tfrac{3}{2},2)-A\mathsf{F}(\tfrac{3}{2},1)\right)-\varepsilon\beta\exp(-2\mathrm{i}\Psi)\left(2\mathsf{F}(\tfrac{3}{2},4)-3A\mathsf{F}(\tfrac{3}{2},3)\right)\\ &-\varepsilon\beta\left(4\mathsf{F}(\tfrac{3}{2},0)+A\mathsf{F}(\tfrac{3}{2},1)\right)+3\varepsilon\beta\exp(-2\mathrm{i}\Psi)\left((2-\tfrac{A^2}{2})\mathsf{F}(\tfrac{5}{2},4)+\frac{A}{2}\mathsf{F}(\tfrac{5}{2},5)-\frac{9}{2}A\mathsf{F}(\tfrac{5}{2},3)+\frac{5}{2}A^2\mathsf{F}(\tfrac{5}{2},2)\right)\\ &+3\varepsilon\beta\left((2+\tfrac{5}{2}A^2)\mathsf{F}(\tfrac{5}{2},0)-4A\mathsf{F}(\tfrac{5}{2},1)-\frac{A^2}{2}\mathsf{F}(\tfrac{5}{2},2)\right)\\ &\simeq \varepsilon\exp(-\mathrm{i}\Psi)\left(1.0397+4.03(a-1)\right)+1.51\varepsilon\beta+5.68\varepsilon\beta\exp(-2\mathrm{i}\Psi)\end{aligned} \tag{7.69a}$$

$$\mathcal{Q}_0 = \oint_{C_0} \mathcal{Q}(q)\frac{\mathrm{d}q}{2\pi\mathrm{i}q} \equiv \frac{1}{2\pi}\int_0^{2\pi} \mathcal{Q}(\xi)\mathrm{d}\xi \simeq$$
$$-2\varepsilon a^2\left(a\mathsf{F}(\tfrac{3}{2},0) - A\mathsf{F}(\tfrac{3}{2},1)\right) - \varepsilon\beta\exp(-\mathrm{i}\Psi)\left(2\mathsf{F}(\tfrac{3}{2},2) - 3A\mathsf{F}(\tfrac{3}{2},1)\right)$$
$$-\varepsilon\beta\exp(\mathrm{i}\Psi)\left(4\mathsf{F}(\tfrac{3}{2},2) + A\mathsf{F}(\tfrac{3}{2},3)\right) + 3\varepsilon\beta\exp(-\mathrm{i}\Psi)$$
$$\times\left((2-\tfrac{A^2}{2})\mathsf{F}(\tfrac{5}{2},2) + \frac{A}{2}\mathsf{F}(\tfrac{5}{2},3) - \tfrac{9}{2}A\mathsf{F}(\tfrac{5}{2},1) + \tfrac{5}{2}A^2\mathsf{F}(\tfrac{5}{2},0)\right)$$
$$+3\varepsilon\beta\exp(\mathrm{i}\Psi)\left((2+\tfrac{5}{2}A^2)\mathsf{F}(\tfrac{5}{2},2) - \tfrac{3}{2}A\mathsf{F}(\tfrac{5}{2},3) - \tfrac{5}{2}A\mathsf{F}(\tfrac{5}{2},1) - \tfrac{A^2}{2}\mathsf{F}(\tfrac{5}{2},4)\right)$$
$$\simeq 0.4390\varepsilon + 1.95\varepsilon(a-1) + 5.07\varepsilon\beta\exp(-\mathrm{i}\Psi) + 1.03\varepsilon\beta\exp(\mathrm{i}\Psi) \tag{7.69b}$$

and

$$\mathcal{Q}_1 = \oint_{C_0} \frac{\mathcal{Q}(q)}{q^2}\frac{\mathrm{d}q}{2\pi\mathrm{i}q} = \frac{1}{2\pi}\int_0^{2\pi} \frac{\mathcal{Q}(\xi)}{\exp(2\mathrm{i}\xi - \mathrm{i}\Psi)}\mathrm{d}\xi$$
$$\simeq -2\varepsilon a^2\exp(\mathrm{i}\Psi)\left(a\mathsf{F}(\tfrac{3}{2},2) - A\mathsf{F}(\tfrac{3}{2},3)\right) - \varepsilon\beta\left(2\mathsf{F}(\tfrac{3}{2},0) - 3A\mathsf{F}(\tfrac{3}{2},1)\right)$$
$$-\varepsilon\beta\exp(2\mathrm{i}\Psi)\left(4\mathsf{F}(\tfrac{5}{2},4) + A\mathsf{F}(\tfrac{5}{2},5)\right) + 3\varepsilon\beta\left((2-\tfrac{A^2}{2})\mathsf{F}(\tfrac{5}{2},0) - 4A\mathsf{F}(\tfrac{5}{2},1) + \tfrac{5}{2}A^2\mathsf{F}(\tfrac{5}{2},2)\right)$$
$$+3\varepsilon\beta\exp(2\mathrm{i}\Psi)\left((2+\tfrac{5}{2}A^2)\mathsf{F}(\tfrac{5}{2},4) - \tfrac{3}{2}A\mathsf{F}(\tfrac{5}{2},5) - \tfrac{5}{2}A\mathsf{F}(\tfrac{5}{2},3) - \tfrac{A^2}{2}\mathsf{F}(\tfrac{5}{2},6)\right)$$
$$\simeq \varepsilon\exp(\mathrm{i}\Psi)\left(0.1191 + 0.79(a-1)\right) + 1.08\varepsilon\beta + 0.53\varepsilon\beta\exp(2\mathrm{i}\Psi) \tag{7.69c}$$

where, similarly to (4.21)

$$\mathsf{F}(\alpha,n) \equiv \frac{a^n}{A^{n+2\alpha}}\frac{\Gamma(\alpha+n)}{n!\Gamma(\alpha)}F(\alpha,\alpha+n,n+1;\tfrac{a^2}{A^2}) \tag{7.70}$$

Note that, in addition to β, we have treated $a-1$ as a small parameter, and expanded our results accordingly (including only linear terms in either β or $a-1$).

It turns out that, to understand the crucial feature of the solution, it is sufficient to quote only the first non-zero term of each time derivative (i.e. occasionally discarding even the $a-1$ and β-proportional terms). Here are the results:

$$a' \simeq 2a\,\mathrm{Im}(\mathcal{Q}_0 - \beta\mathcal{Q}_{-1}) \simeq -6.00\varepsilon\beta\sin(\Psi) \tag{7.71}$$

$$\beta' \simeq -\frac{\mathrm{Im}(\mathcal{Q}_1 + 3\mathcal{Q}_{-1})}{4} \simeq 0.75\varepsilon\sin(\Psi) \tag{7.72}$$

and

$$\Psi' = 2\psi' - 2s'_{\mathrm{p}} - 2(a^{3/2}-1) \simeq -\frac{\mathrm{Re}(\mathcal{Q}_1 - 3\mathcal{Q}_{-1})}{4\beta} - 3(a-1) \simeq \frac{0.75\varepsilon\cos(\Psi)}{\beta} - 3(a-1) \tag{7.73}$$

In the last equation, it was necessary to keep two terms (one proportional to ε/β, the other to $a-1$), since they are of similar magnitude. One can show that including additional terms would not change the qualitative nature of the solution (even the quantitative changes are rather minute).

The first two equations imply that $4\beta^2 + a$ must be a constant, say $2K+1$. Similarly, the last two equations (with a replaced by $2K+1-4\beta^2$) imply that

$$0.75\varepsilon\beta\cos(\Psi) - 3\beta^2K + 3\beta^4 \tag{7.74}$$

is another constant.

Proof.

$$
\begin{aligned}
&0.75\varepsilon\beta'\cos(\Psi) - 0.75\varepsilon\beta\sin(\Psi)\Psi' - 6\beta K\beta' + 12\beta^3\beta' \\
&= \left(0.75\varepsilon\cos(\Psi) - 6\beta K + 12\beta^3\right)0.75\varepsilon\sin(\Psi) - 0.75\varepsilon\beta\sin(\Psi)\left(\frac{0.75\varepsilon\cos(\Psi)}{\beta} - 6K + 12\beta^2\right) \\
&= 0
\end{aligned}
\tag{7.75}
$$

□

The contours of (7.74) thus yield a family of solutions to our differential equations.

Note that the fixed points are obtained by solving

$$\sin\Psi = 0 \tag{7.76a}$$

$$0.75\varepsilon\cos(\Psi) - 6K\beta + 12\beta^3 = 0 \tag{7.76b}$$

This yields two 'physical' (i.e. $0 < \beta < 1$) solutions at $\Psi = 0$ when the discriminant of the cubic equation, namely

$$\left(-\frac{K}{6}\right)^3 + \left(\frac{0.75}{24}\varepsilon\right)^2 \tag{7.77}$$

is negative (i.e. when $K > 0.00595$), and another one at $\Psi = \pi$ (for any K). We denote the corresponding values of β by subscripts 1 (the smaller β) 2 (the bigger one) and 3, respectively.

By analyzing the corresponding linearized version of the two equations, which reads

$$(\beta - \beta_k)' = 0.75\varepsilon\Psi \tag{7.78a}$$

$$\Psi' = \left(-\frac{0.75\varepsilon}{\beta_k^2} + 24\beta_k\right)(\beta - \beta_k) \tag{7.78b}$$

for $k = 1$ or 2, and

$$(\beta - \beta_3)' = -0.75\varepsilon(\Psi - \pi) \tag{7.78c}$$

$$(\Psi - \pi)' = \left(\frac{0.75\varepsilon}{\beta_3^2} + 24\beta_3\right)(\beta - \beta_3) \tag{7.78d}$$

for the last fixed point, one discovers that β_1 and β_3 correspond to two centers, and β_2 is a saddle point. Using $K = 0.008$ as an example, we display the corresponding family of solutions in Figure 7.4.

When $K \le 0.00595$, the first two fixed points disappear, and only the the one at $\Psi = \pi$ is left. The new set of solutions is illustrated in Figure 7.5 (taking $K = 0.00595$).

Secondly, we show (in Figure 7.6) how the corresponding equilibrium values of a vary with K (as we already know, we have two of these for $K > 0.00595$ and only one when $K \le 0.00595$), and what the corresponding frequencies (relative to the asteroid's orbital frequency) are (Figure 7.7).

7.2.2 Gap formation

Our analysis has shown so far that, at or near the 2/1 resonance, asteroids' semi-major axis is forced to undergo large (and relatively slow) oscillations. But, that by itself does *not* lead to a gap creation! The situation markedly changes when we add other perturbing forces, such as Kepler shear introduced in Chapter 3. One can show that this adds $-11.06\beta^3\rho C/m$ and $-3.08\beta^2\rho C/m$ to the right hand side of (7.71) and (7.72), respectively, where m is *ordinary* mass of the asteroid. This means that K is no longer constant, but it slowly decreases (at the rate of $-11.7\beta^3\rho C/m$). Most solutions of the $K > 0.00595$ type

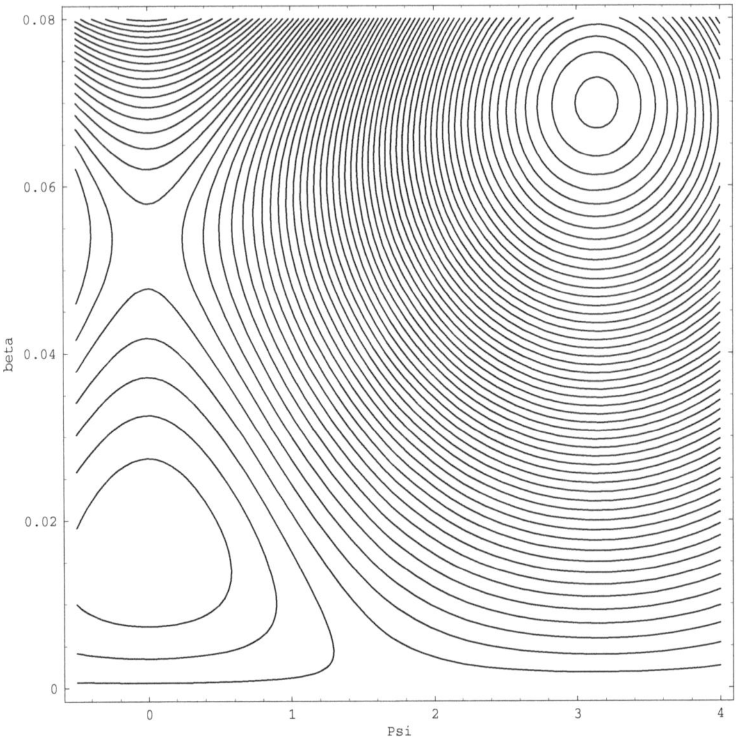

Figure 7.4 Solutions near 2/1 resonance ($K = 0.00800$).

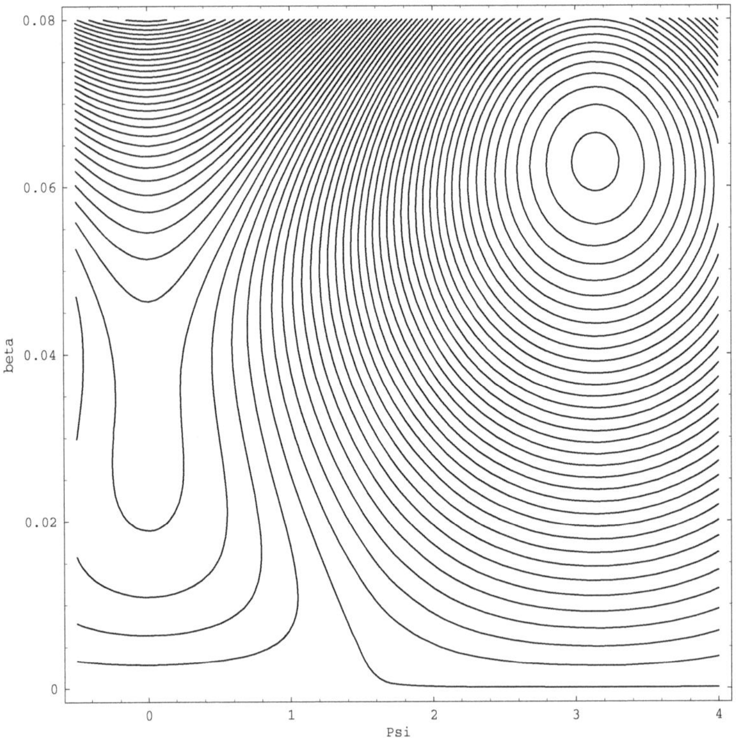

Figure 7.5 Solutions near 2/1 resonance ($K = 0.00595$).

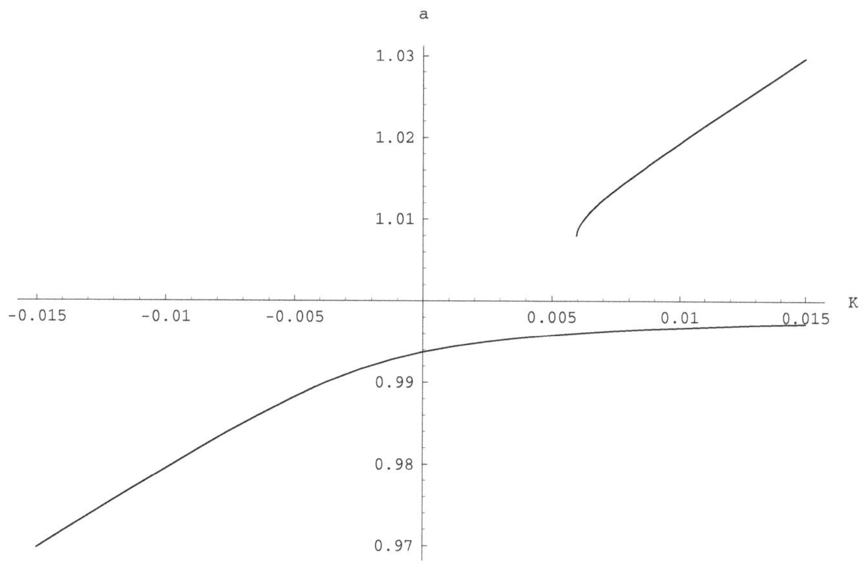

Figure 7.6 Semi-major axis at equilibrium.

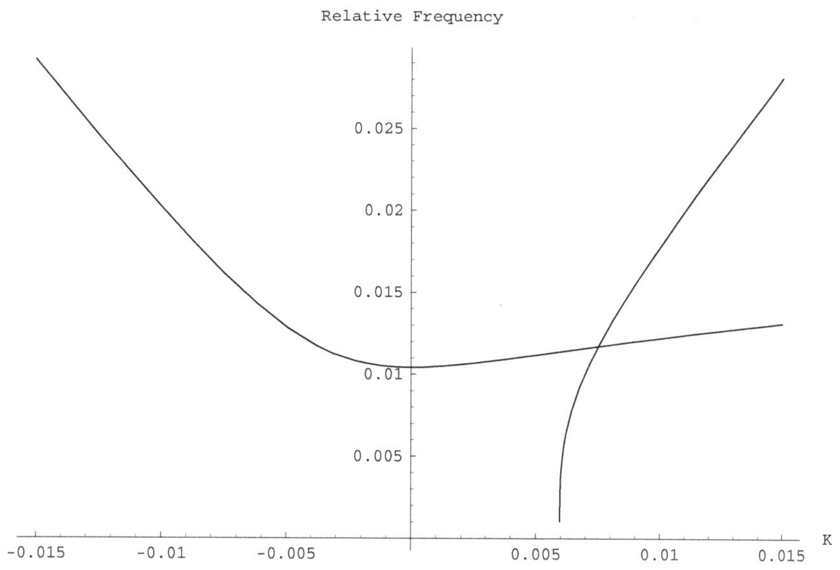

Figure 7.7 Frequency of small oscillations around equilibrium.

must then, sooner or later, qualitatively change, when one of its two centers suddenly disappears. The corresponding effect on the vale of a is rather dramatic, as a numerical solution of the new set of equations (using $\rho C/m \simeq 0.0002$) clearly demonstrates (see Figure 7.8). Creation of a gap in the $a \simeq 1$ region is then

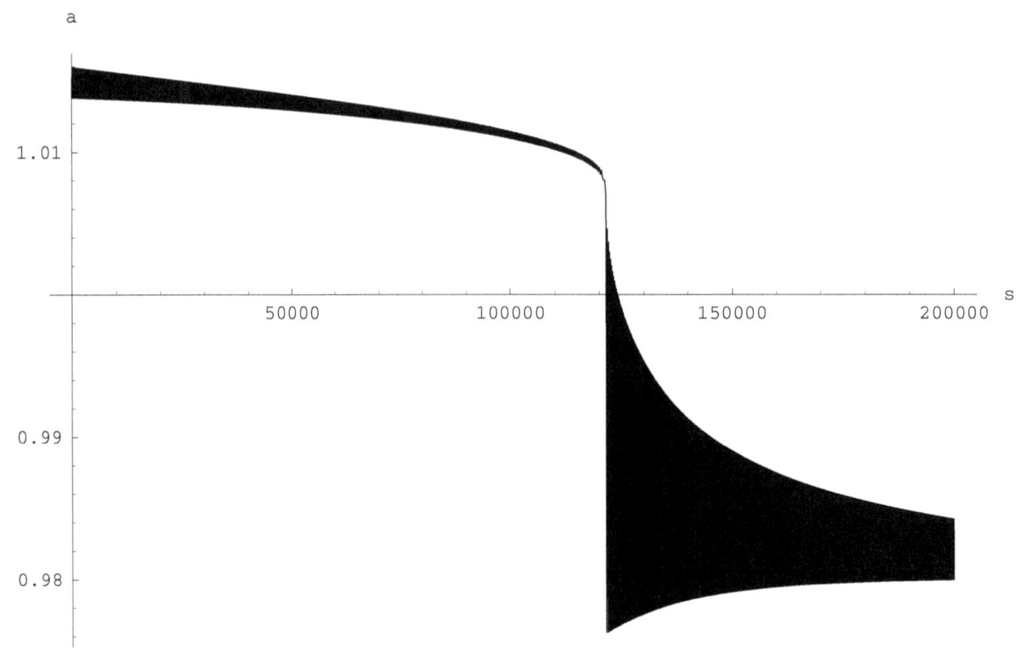

Figure 7.8 Gap formation (restricted problem).

quite obvious.

The same mechanism is responsible for the so called CASSINI DIVISION (a very sharp and well defined gap) in Saturn's ring.

7.2.3 Elliptical perturbing orbit

We now assume that the perturbing orbit has a small eccentricity γ, and compute its contribution to our differential equations, to the first-order accuracy.

Based on (6.3), $\mathbf{R}_o$, $\mathbf{r}^*\mathbf{R}_o$, R_o^2 and $|\mathbf{r}_o - \mathbf{R}_o|^{-3}$ acquire an extra

$$\cdots + \gamma A \frac{z \exp(-i\Psi + i\psi) - 3\exp(-i\psi)}{2} \tag{7.79a}$$

$$\cdots + \gamma A \frac{\exp(-i\Psi + i\psi) - 3\exp(i\Psi - i\psi - 2i\xi)}{2} \tag{7.79b}$$

$$\cdots - \gamma A^2 \left(\exp(i\xi - i\Psi + i\psi) + \exp(-i\xi + i\Psi - i\psi)\right) \tag{7.79c}$$

and

$$\cdots - \frac{\frac{3}{4}\gamma A \begin{pmatrix} 3\exp(-2i\xi + i\Psi - i\psi) + 3\exp(2i\xi - i\Psi + i\psi) - \exp(i\Psi - i\psi) - \\ \exp(-i\Psi + i\psi) + 2A\exp(-i\xi + i\Psi - i\psi) + 2A\exp(i\xi - i\Psi + i\psi) \end{pmatrix}}{(A^2 + 1 - 2A\cos\xi)^{5/2}} \tag{7.80}$$

respectively (the last numerator has been broken into two lines, due to its length).

This implies that $\mathcal{Q}(z)$ (see Eq. 7.38) must be extended by

$$\varepsilon\gamma A\left(\exp(-\mathrm{i}\Psi+\mathrm{i}\psi)-3\exp(\mathrm{i}\Psi-\mathrm{i}\psi-2\mathrm{i}\xi)\right)\left(\frac{1}{(A^2+1-2A\cos\xi)^{3/2}}-1\right)$$
$$+\frac{3}{2}\varepsilon\gamma A\left(1-A\exp(-\mathrm{i}\xi)\right)\frac{\begin{pmatrix}3\exp(-2\mathrm{i}\xi+\mathrm{i}\Psi-\mathrm{i}\psi)+3\exp(2\mathrm{i}\xi-\mathrm{i}\Psi+\mathrm{i}\psi)-\exp(\mathrm{i}\Psi-\mathrm{i}\psi)-\\ \exp(-\mathrm{i}\Psi+\mathrm{i}\psi)+2A\exp(-\mathrm{i}\xi+\mathrm{i}\Psi-\mathrm{i}\psi)+2A\exp(\mathrm{i}\xi-\mathrm{i}\Psi+\mathrm{i}\psi)\end{pmatrix}}{(A^2+1-2A\cos\xi)^{5/2}}$$
$$+3\varepsilon\gamma A^3\exp(-\mathrm{i}\xi)\left(\exp(\mathrm{i}\xi-\mathrm{i}\Psi+\mathrm{i}\psi)+\exp(-\mathrm{i}\xi+\mathrm{i}\Psi-\mathrm{i}\psi)\right) \tag{7.81}$$

leading to (see Eqs. 7.69a, 7.69b and 7.69c)

$$\mathcal{Q}_{-1}\simeq\cdots+\varepsilon\gamma\left(0.615\exp(-\mathrm{i}\psi)+4.325\exp(-2\mathrm{i}\Psi+\mathrm{i}\psi)\right) \tag{7.82a}$$
$$\mathcal{Q}_{0}\simeq\cdots+\varepsilon\gamma\left(1.375\exp(-2\mathrm{i}\Psi+\mathrm{i}\psi)+0.835\exp(\mathrm{i}\Psi-\mathrm{i}\psi)\right) \tag{7.82b}$$
$$\mathcal{Q}_{-1}\simeq\cdots+\varepsilon\gamma\left(0.395\exp(\mathrm{i}\psi)+0.459\exp(2\mathrm{i}\Psi-\mathrm{i}\psi)\right) \tag{7.82c}$$

where three dots represent the old terms.

Correspondingly extending (7.71)-(7.73), one gets

$$a'\simeq\cdots-1.079\varepsilon\gamma\sin(\Psi-\psi) \tag{7.83a}$$
$$\beta'\simeq\cdots+0.363\varepsilon\gamma\sin\psi+3.129\varepsilon\gamma\sin(2\Psi-\psi) \tag{7.83b}$$
$$\Psi'\simeq\cdots+\frac{0.363\varepsilon\gamma\cos\psi+3.129\varepsilon\gamma\cos(2\Psi-\psi)}{\beta} \tag{7.83c}$$

and the following equation for the extra variable ψ (angular distance between the heavy and massless satellites' apocenters).

$$\psi'\simeq-\frac{\mathrm{Re}(\mathcal{Q}_1-3\mathcal{Q}_{-1})}{4\beta}=\frac{0.75\varepsilon\cos(\Psi)+0.363\varepsilon\gamma\cos\psi+3.129\varepsilon\gamma\cos(2\Psi-\psi)}{\beta} \tag{7.83d}$$

Surprisingly, these extra few (and relatively small) terms change the nature of the equations, again rather dramatically. The solutions are no longer periodic; furthermore, most of them exhibit an extreme sensitivity to initial conditions (such a behavior is called 'CHAOTIC'). To demonstrate this, in Figure 7.9 we display results of numerical integration of the new set of equations (without Kepler shear) in terms of β.

Note that another aspect of this chaotic behavior is to force eccentricity, occasionally, into fairly high values — in our previous example, this was about $\beta = 0.6$ (0.88, in terms of *regular* eccentricity). Yet, this is definitely not the mechanism of gap clearing (as sometimes incorrectly claimed), which can be seen from the corresponding behavior of a, shown in Figure 7.10.

7.2.4 Nonzero inclination

When the asteroid has a nonzero inclination with respect to the perturbing orbit, ψ in all formulas of the previous two sections needs to be replaced by $\psi+\phi$, e.g.

$$\Psi\equiv2\psi+2\phi-2s_{\mathrm{p}}-2\int(a^{3/2}-1)\mathrm{d}s \tag{7.84a}$$
$$a'\simeq-6\varepsilon\beta\sin(\Psi)-1.079\varepsilon\gamma\sin(\Psi-\psi-\phi) \tag{7.84b}$$

etc. (ψ alone now represents the angular distance of the asteroid's perihelion from its node).

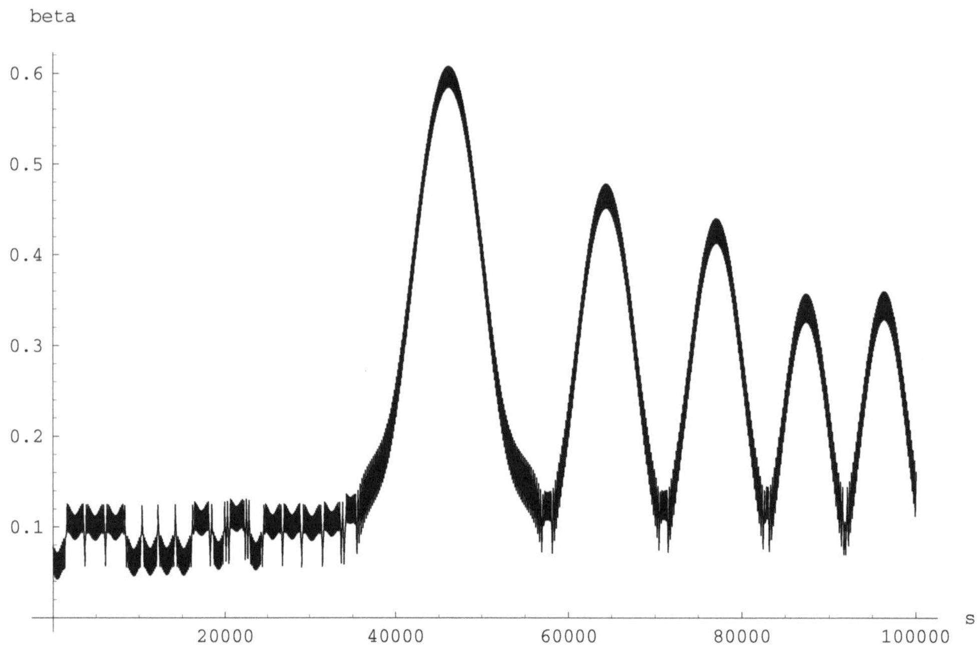

Figure 7.9 Asteroid's eccentricity variation (Jupiter's eccentricity = 0.05).

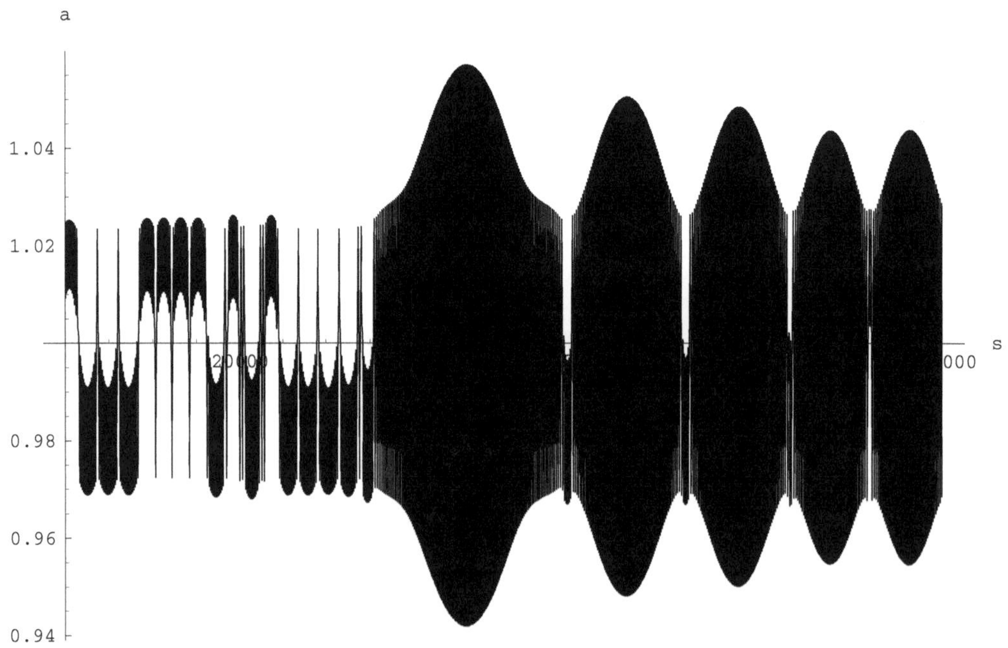

Figure 7.10 Variation of semi-major axis (Jupiter's eccentricity = 0.05).

We then have to revert back to the quaternion formulation of the problem. Since first-degree terms in θ do not contribute to our equations, we have to carry out all expansions to θ^2. This results in

$$\mathbf{R} \simeq A\mathfrak{k}\exp(-\mathrm{i}\frac{\Psi}{2})q\left(1+\frac{\beta}{2}z-\frac{\beta}{2z}-\frac{\theta^2}{4}\right)+A\mathfrak{k}\frac{\exp(\mathrm{i}\frac{\Psi}{2}-2\mathrm{i}\psi)}{4q}\theta^2+\frac{A}{2}\left(\frac{\exp(\mathrm{i}\frac{\Psi}{2}-\mathrm{i}\psi)}{q}-\exp(-\mathrm{i}\frac{\Psi}{2}+\mathrm{i}\psi)q\right)\theta$$
$$+\frac{A}{4}\frac{a^{3/2}}{q}(1-z^2)\left(\frac{\exp(\mathrm{i}\frac{\Psi}{2}-\mathrm{i}\psi)}{z}+\exp(-\mathrm{i}\frac{\Psi}{2}+\mathrm{i}\psi)z\right)\beta\theta+\cdots \tag{7.85}$$

from which one can easily derive $\overline{\mathbf{r}_o}\mathbf{R}_o$, $|\mathbf{r}_o-\mathbf{R}_o|^{-3}$ (note that $R_o^2=A^2$, since, for this computation, we can go back to $\gamma=0$), $\mathcal{Q}(z)$ and $\mathcal{W}(z)$. This time we skip the (rather tedious) details of deriving the individual coefficients of the last two quantities, and proceed to quote the final results:

$$a' \simeq \cdots - 1.033\varepsilon\theta^2\sin(2\Psi-2\psi) \tag{7.86a}$$
$$\beta' \simeq \cdots + \varepsilon\theta^2\left[0.497\sin(3\Psi-2\psi)-0.551\sin(\Psi-2\psi)-1.704\sin(\Psi)\right] \tag{7.86b}$$
$$\Psi' \simeq \cdots + \varepsilon\theta^2\frac{0.497\cos(3\Psi-2\psi)+0.551\cos(\Psi-2\psi)-1.704\cos(\Psi)}{\beta} \tag{7.86c}$$
$$(\phi+\psi)' \simeq \cdots + \varepsilon\theta^2\frac{0.497\cos(3\Psi-2\psi)+0.551\cos(\Psi-2\psi)-1.704\cos(\Psi)}{\beta} \tag{7.86d}$$
$$\psi' \simeq (\phi+\psi)' \tag{7.86e}$$

and an extra

$$\theta' \simeq 0.516\varepsilon\theta\sin(2\Psi-2\psi) \tag{7.87}$$

With the exception of the last equation, the leading θ-terms are always of the second degree. Their effect is similar (but weaker) to that of γ (namely making the originally well-behaved set of solutions chaotic).

7.3 HIGHER-ORDER RESONANCES

We now present the corresponding results for selected higher-order resonances [48]. As a rule, we always quote only the leading non-zero term in each of the β, γ and θ expansions.

7.3.1 3/2 resonance

$$\frac{a'}{\varepsilon} \simeq -24.73\beta\sin(\Psi)-15.17\gamma\sin(\Psi-\phi-\psi)-7\theta^2\sin(2\Psi-2\psi) \tag{7.88a}$$
$$\frac{\beta'}{\varepsilon} \simeq 1.546\sin(\Psi)+1.527\gamma\sin(\phi+\psi)+10.15\gamma\sin(2\Psi-\phi-\psi) \tag{7.88b}$$
$$+\theta^2\left(3.170\sin(3\Psi-2\psi)-2.342\sin(\Psi-2\psi)-9.631\sin(\Psi)\right) \tag{7.88c}$$
$$\frac{(\phi+\psi)'}{\varepsilon} \simeq \frac{\psi'}{\varepsilon} \simeq \frac{1.546\cos(\Psi)+1.527\gamma\cos(\phi+\psi)+10.15\gamma\cos(2\Psi-\phi-\psi)}{\beta} \tag{7.88d}$$
$$+\theta^2\frac{3.170\cos(3\Psi-2\psi)+2.342\cos(\Psi-2\psi)-9.631\cos(\Psi)}{\beta} \tag{7.88e}$$
$$\Psi' \simeq (\phi+\psi)'-6(a-1) \tag{7.88f}$$
$$\frac{\theta'}{\varepsilon} \simeq -1.75\theta\sin(2\Psi-2\psi) \tag{7.88g}$$

where

$$\Psi \equiv 3\psi + 3\phi - 4s_{\mathrm{p}} - 4\int (a^{3/2} - 1)\mathrm{d}s \tag{7.89}$$

One can thus see that the results are quite similar to the 2/1 case (same terms, with different coefficients). This appears to be the situation for any $r/(r-1)$ resonance.

7.3.2 3/1 resonance[2]

$$\frac{a'}{\varepsilon} \simeq -4.606\beta^2 \sin(\Psi) - 8.511\beta\gamma \sin(\Psi - \phi - \psi) - 0.159\theta^2 \sin(\Psi - 2\psi) \tag{7.90a}$$

$$\frac{\beta'}{\varepsilon} \simeq 1.151\beta \sin(\Psi) + 0.080\gamma \sin(\phi + \psi) + 1.064\gamma \sin(\Psi - \phi - \psi) \tag{7.90b}$$

$$\frac{(\phi+\psi)'}{\varepsilon} \simeq \frac{\psi'}{\varepsilon} \simeq 0.274 + 1.151\cos(\Psi) + 0.080\gamma\frac{\cos(\phi+\psi)}{\beta} + 1.064\gamma\frac{\cos(\Psi - \phi - \psi)}{\beta} \tag{7.90c}$$

$$\Psi' \simeq 2(\phi+\psi)' - 0.075\varepsilon - 3(a-1) \tag{7.90d}$$

$$\frac{\theta'}{\varepsilon} \simeq -0.159\theta \sin(\Psi - 2\psi) \tag{7.90e}$$

where

$$\Psi \equiv 3\psi + 3\phi - 2s_{\mathrm{p}} - 2\int (a^{3/2} - 1)\mathrm{d}s \tag{7.91}$$

Here, the γ and θ^2-proportional terms are of the same magnitude as the 'regular' (no γ or θ) terms of the equations, resulting in a much more pronounced chaotic behavior.

7.3.3 5/2 resonance

$$\frac{a'}{\varepsilon} \simeq 39.36\beta^3 \sin(\Psi) - 98.78\gamma\beta^2 \sin(\Psi - \psi - \phi) - 2.799\beta\theta^2 \sin(\Psi - 2\psi) - 2.703\gamma\theta^2 \sin(\Psi - 3\psi - \phi) \tag{7.92a}$$

$$\frac{\beta'}{\varepsilon} \simeq 0.151\gamma \sin(\phi + \psi) + 7.380\beta^2 \sin(\Psi) + 0.175\theta^2 \sin(\Psi - 2\psi) \tag{7.92b}$$

$$\frac{\psi'}{\varepsilon} \simeq 0.501\gamma\frac{\cos(\phi+\psi)}{\beta} + 7.380\beta\cos(\Psi) + 1.399\beta\cos(\Psi - 2\psi) + 0.175\theta^2\frac{\cos(\Psi - 2\psi)}{\beta} \tag{7.92c}$$

$$\frac{(\phi+\psi)'}{\varepsilon} \simeq 0.464 + 0.501\gamma\frac{\cos(\phi+\psi)}{\beta} + 7.380\beta\cos(\Psi) + 0.175\theta^2\frac{\cos(\Psi - 2\psi)}{\beta} \tag{7.92d}$$

$$\Psi' \simeq 1.156\varepsilon + 0.453\varepsilon\gamma\frac{\cos(\phi+\psi)}{\beta} + 22.14\varepsilon\beta\cos(\Psi) + 0.525\varepsilon\theta^2\frac{\cos(\Psi - 2\psi)}{\beta} - 6(a-1) \tag{7.92e}$$

$$\frac{\theta'}{\varepsilon} \simeq -1.399\theta\beta \sin(\Psi - 2\psi) - 1.352\theta\gamma \sin(\Psi - 3\psi - \phi) \tag{7.92f}$$

where

$$\Psi \equiv 5\psi + 5\phi - 4s_{\mathrm{p}} - 4\int (a^{3/2} - 1)\mathrm{d}s \tag{7.93}$$

In this case, the γ-proportional terms become dominant.

---

[2]See [57].

7.3.4 7/3 resonance

To simplify the formulas (which in this case would be getting rather lengthy) we now assume that the orbit has zero inclination.

$$\frac{a'}{\varepsilon} \simeq 244.9\beta^4 \sin(\Psi) - 788.9\beta^3\gamma \sin(\Psi - \phi - \psi) \tag{7.94a}$$

$$\frac{\beta'}{\varepsilon} \simeq \gamma\left(0.195 + 2.407\beta^2\right)\sin(\phi + \psi) + 40.82\beta^3 \sin(\Psi) + 98.61\gamma\beta^2 \sin(\Psi - \phi - \psi) \tag{7.94b}$$

$$\frac{(\phi+\psi)'}{\varepsilon} \simeq 0.195\gamma\frac{\cos(\phi+\psi)}{\beta} + 0.576 + 1.886\beta^2 + 40.82\beta^2\cos(\Psi) + 6.439\beta\gamma\cos(\phi+\psi) + 98.61\beta\gamma\cos(\Psi - \phi - \psi) \tag{7.94c}$$

$$\Psi' \simeq 1.878\varepsilon + 0.782\varepsilon\gamma\frac{\cos(\phi+\psi)}{\beta} + 4.403\varepsilon\beta^2 + 163.3\varepsilon\beta^2\cos(\Psi) + 21.41\varepsilon\beta\gamma\cos(\phi+\psi) + 394.4\varepsilon\beta\gamma\cos(\Psi-\phi-\psi) - 9(a-1) \tag{7.94d}$$

where

$$\Psi \equiv 7\psi + 7\phi - 6s_{\mathrm{p}} - 6\int (a^{3/2} - 1)\mathrm{d}s. \tag{7.95}$$

Corresponding to a general pattern (which clearly emerges at this point), the importance of β-related terms has further decreased (by acquiring an extra power of β).

To demonstrate a gap creation at this resonance as well, we assume that both θ and γ are zero (chaotic terms tend to obscure the underlying picture), but recognize that there will be other extra small perturbations (from the remaining planets — mainly Saturn, etc.). One can show that, whatever the new perturbation is, its effect is always similar to that of Kepler shear (in terms of gap creation). Thus, for example, if we consider a perturbation which would subtract a small term, say 0.000000001, from the right hand side of (7.94a), we get Figure 7.11.

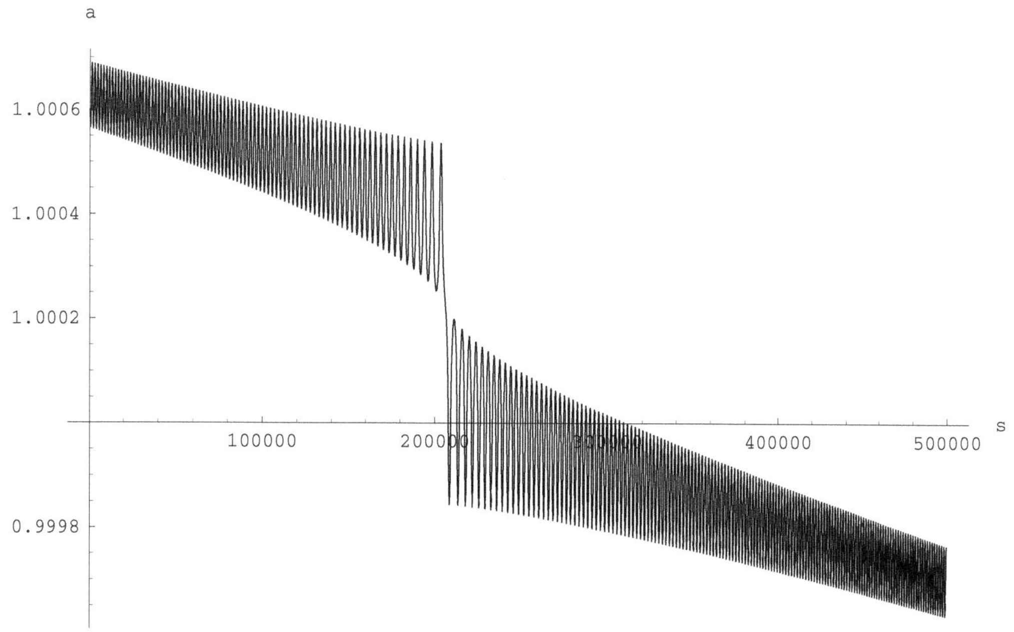

Figure 7.11 Gap formation at 7/3 resonance with a small negative perturbation.

Clearly, no asteroid is allowed to spend much time in the $a \simeq 1$ region (which now corresponds to the exact 7/3 resonance), even though the resulting gap is not as wide as it was in the 2/1 case.

Similarly, just by reversing the sign of the new perturbing term, we get Figure 7.12.

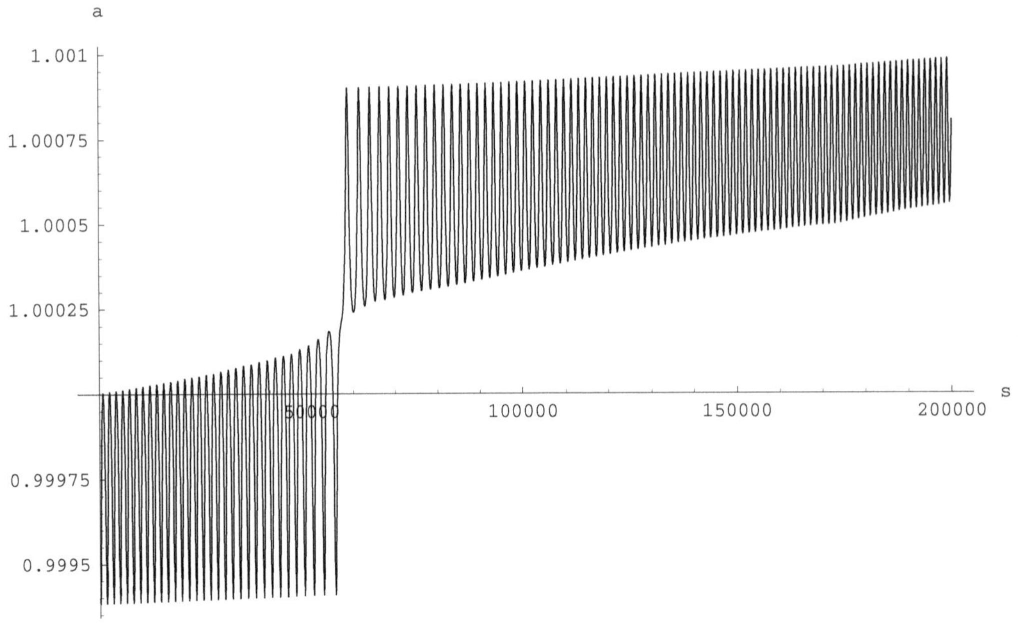

Figure 7.12 Gap formation at 7/3 resonance (positive perturbation).

One can thus see that a gap is created in either situation.

CHAPTER 8

Other Perturbations

Abstract

In this chapter we briefly discuss most of the remaining perturbing forces. None of them have an explicit time dependence, which simplifies the resulting differential equations for the orbital elements (making them autonomous). Furthermore, the results are developed only to the first-order-in-ε accuracy. With the exception of the distant-source radiation, all perturbing forces discussed here act within the orbital plane; we can thus treat all quantities as complex (no quaternions needed) — this also eliminates θ and ϕ from our orbital-element list. Finally (with the same single exception), the perturbing forces involve no special, fixed direction, which eliminates ψ from the right hand side of the resulting differential equations, making them functions of a and β only.

8.1 RELATIVISTIC CORRECTIONS

Rewriting the perturbing force (3.65) in Kepler's frame as

$$\varepsilon \mathbf{f}_\text{o} = \varepsilon \left(4\frac{\mathbf{r}_\text{o}}{r^4} + \mathbf{r}'_\text{o}\frac{\mathbf{r}'_\text{o}\cdot\mathbf{r}_\text{o}}{ar^5} - \mathbf{r}_\text{o}\frac{\mathbf{r}'_\text{o}\cdot\mathbf{r}'_\text{o}}{4ar^5} \right) \tag{8.1}$$

(where $\varepsilon \equiv \mu^2/c^2$ meter^4 sec^{-2}) we can compute $\mathcal{Q}(z)$ (based on the first line of Eq. 5.7), using (5.3) and (5.4).

This results in

$$\mathcal{Q}(z) = -\frac{2\varepsilon}{a\mu}\frac{(1+\beta^2)\left(3\frac{\beta^2}{z^2} - \frac{\beta}{z}(8+3\beta^2) - (3+8\beta^2) + 3\beta z\right)}{(1+\frac{\beta}{z})^3(1+\beta z)^3} \tag{8.2}$$

since, to a sufficient $O(\varepsilon^0)$ accuracy

$$\mathbf{r}_\text{o} = \mathfrak{k}\frac{a(z+\beta)^2}{z(1+\beta^2)} \tag{8.3a}$$

$$\mathbf{r}'_\text{o} = 2\mathfrak{j}\frac{a(z^2-\beta^2)}{z(1+\beta^2)} \tag{8.3b}$$

$$r = \frac{a}{1+\beta^2}(1+\frac{\beta}{z})(1+\beta z) \tag{8.3c}$$

and

$$\mathbf{r}'_\text{o}\cdot\mathbf{r}_\text{o} = \frac{2\mathrm{i}a^2\beta(z-\frac{1}{z})(1+\frac{\beta}{z})(1+\beta z)}{(1+\beta^2)^2} \tag{8.4}$$

$$\mathbf{r}'_\text{o}\cdot\mathbf{r}'_\text{o} = \frac{4a^2(1-\frac{\beta^2}{z^2})(1-\beta^2 z^2)}{(1+\beta^2)^2} \tag{8.5}$$

At the same time, $\mathcal{W}(z)$ is equal to zero - this is essentially a planar problem.

Jan Vrbik

After evaluating a few contour integrals, we get

$$a' = 0 \tag{8.6a}$$

$$\beta' = 0 \tag{8.6b}$$

$$\psi' = \frac{6\varepsilon}{a\mu}\left(\frac{1+\beta^2}{1-\beta^2}\right)^2 \equiv \frac{6\mu}{ac^2(1-e^2)} \tag{8.6c}$$

where e is the ordinary eccentricity. To convert the last quantity to arcseconds per *year*, one has to multiply it by $\frac{360}{2\pi} \times 60^2$ (to convert from radians to arcseconds) and divide by $\frac{T}{\pi}$ (to adjust the time unit) where T is the planet's orbital period in years. For Mercury, the formula yields the 'missing' 43″ per *century*. For the other planets, the effect quickly decreases with their distance from Sun (being proportional to $a^{-5/2}$); e.g. for Earth, we get only 3.8″ per century.

Similarly, the remaining results are

$$s_{\mathrm{p}}' = \frac{\mu}{ac^2}\frac{9+15\beta^2}{2(1-\beta^2)} \tag{8.7}$$

and

$$\mathbf{r}_{\mathrm{o}} = \mathfrak{k}\left[\frac{a(z+\beta)^2}{z(1+\beta^2)} - \frac{\mu}{c^2}\frac{\beta(1+\frac{\beta}{z})(1-7\beta^2+7z\beta-z\beta^3)}{(1-\beta^2)^2} + \frac{\mu}{c^2}\frac{(1+\frac{\beta}{z})(z-7\beta-7z\beta^2+\beta^3)}{(1-\beta^2)^2}\ln\left(\frac{1+\beta z}{1+\frac{\beta}{z}}\right)\right] \tag{8.8}$$

This can be simplified to

$$\mathbf{r}_{\mathrm{o}} = \mathfrak{k}\left(\frac{a(z+\beta)^2}{z(1+\beta^2)} + \frac{\mu}{c^2}z^2\beta + \frac{\mu}{c^2}\frac{15+2z^2-z^4}{2z}\beta^2 + \cdots\right) \tag{8.9}$$

when the μ/c^2 proportional part of the expression is expanded in β (note that the corresponding series is slow-converging, and should be used only for near-circular orbits).

To demonstrate the corresponding orbit's distortion, we plot both (8.8) and (8.9) in Figure 8.1, together with the unperturbed ellipse (regular, dashed and thick line, respectively), using $\beta = 0.5$ and $\mu/c^2 = 0.05$ (to magnify the effect, we have chosen atypically large values):

Note that the orbit has been drawn in Kepler's frame, which itself experiences steady rotation.

8.2 KEPLER SHEAR

This time, using (3.58), the Kepler-frame perturbing force is

$$\varepsilon\mathbf{f}_{\mathrm{o}} = \varepsilon\frac{\mu\beta^2}{4a}\mathrm{j}\sqrt{10-3z^2-\frac{3}{z^2}}\left(3-z^2-\frac{3-2z^2-13z^4+8z^6}{1-3z^2}\beta+\cdots\right) \tag{8.10}$$

where $\varepsilon = \rho C/m$ (of dimension meter^{-1}). This can be easily converted to $\mathcal{Q}(z)$, and after a few contour integrals of the (2.32) type (this time, evaluated by replacing z by $\exp(\mathrm{i}\omega)$, and integrating over ω), we get

$$a' = 2a\mathrm{Im}\left[\oint_{C_0}(1-\beta e^{\mathrm{i}\omega})\mathcal{Q}\frac{\mathrm{d}\omega}{2\pi}\right] = -11.06\varepsilon a^2\beta^3 + O(\beta^5) \tag{8.11a}$$

$$\beta' = -3.08\varepsilon a\beta^2 + O(\beta^4) \tag{8.11b}$$

$$\psi' = 0 \tag{8.11c}$$

$$s_{\mathrm{p}}' = 0 \tag{8.11d}$$

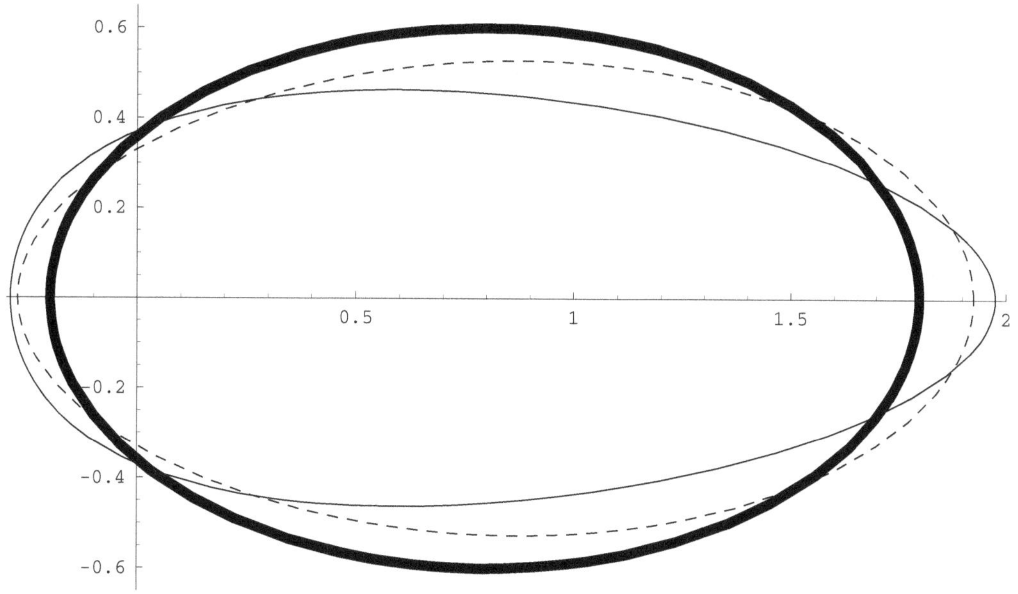

Figure 8.1 Relativistic orbit's distortion.

where the decimal coefficients are, in their exact form, combinations of elliptic integrals.

One can show that, to a sufficient accuracy, the solution to these equations is

$$\beta = \left(\frac{1}{\beta_0} + 3.08\varepsilon a_0 s\right)^{-1} \tag{8.12}$$

$$a = a_0 \left(1 + \frac{11.06\beta_0^2}{2 \times 3.08} - \frac{11.06\beta^2}{2 \times 3.08}\right)^{-1} \tag{8.13}$$

where a_0 and β_0 are the initial values, and s is modified time. Note that β tends to zero, whereas a decreases its value by a relatively small amount of about $1.8\beta_0^2 a_0$ (practically independent of ε). At the same time, the semi-*minor* axis actually increases from its original value of

$$a_0 \frac{1 - \beta_0^2}{1 + \beta_0^2} \simeq a_0(1 - 2\beta_0^2) \tag{8.14}$$

to about $a_0(1 - 1.8\beta_0^2)$. This can be observed in Figure 8.2, which displays the (numerically) exact solution, using $\varepsilon = 0.01$.

In terms of orbit distortion, we get

$$\mathbf{r}_o = \mathfrak{k}\left(\frac{a(z+\beta)^2}{z(1+\beta^2)} + \mathrm{i}\varepsilon\beta^2 a^2 \left(\cdots + 0.07z^{-2} + 0.89 - 0.01z^4 + \cdots\right)\right) \tag{8.15}$$

which was expanded not only in β, but also in powers of z (as the coefficients quickly decrease towards higher powers — both negative and positive). Using exaggerated values of $\beta = 0.5$ and $\varepsilon = 0.2$, the new orbit (thin line) is illustrated in Figure 8.3.

Similarly to the distortion of Moon's orbit, this result is somehow counter-intuitive.

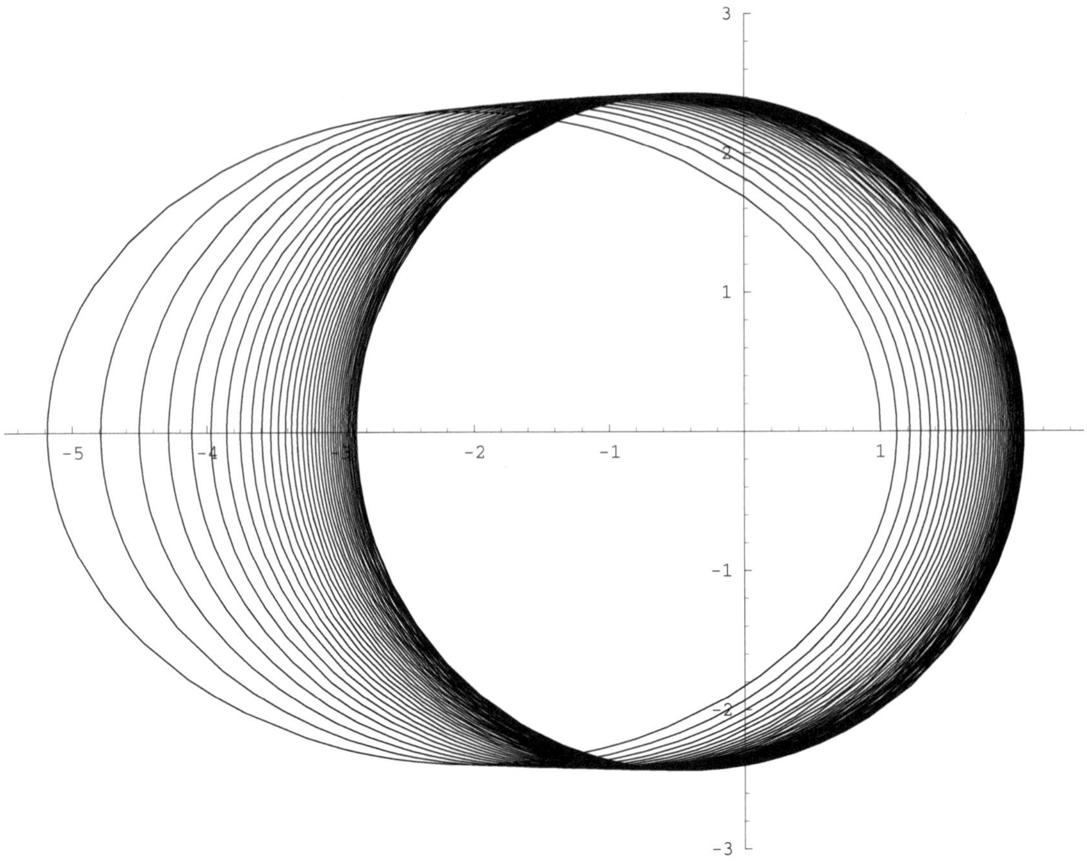

Figure 8.2 Satellite's motion perturbed by Kepler shear.

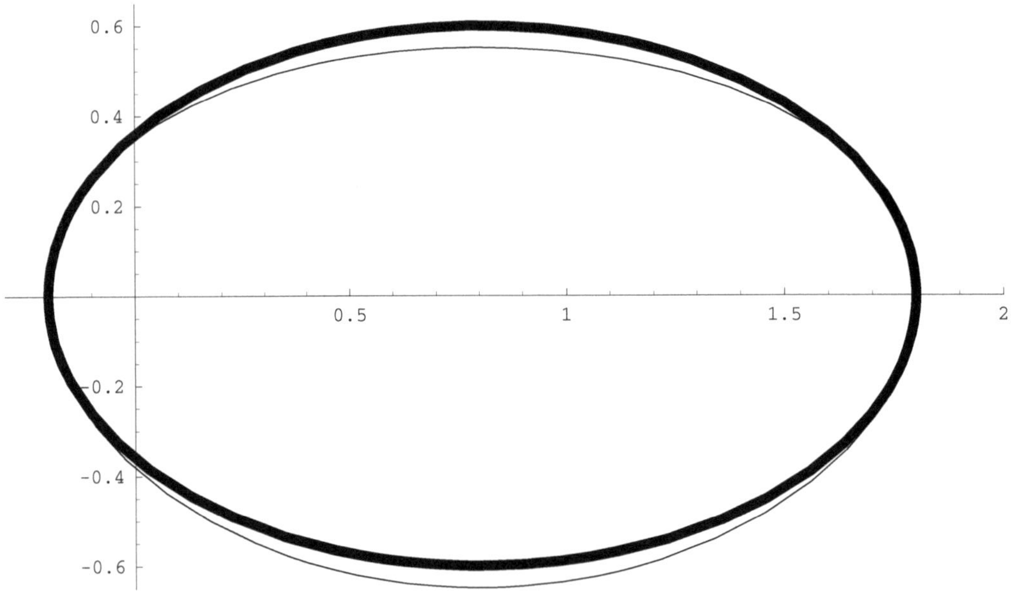

Figure 8.3 Single orbit's distortion due to Kepler shear.

8.3 DRAG

For the perturbing force we use (3.55), assuming that the air density decreases exponentially with altitude, i.e.

$$\rho(r) = \rho_0 e^{-\eta r} \tag{8.16}$$

To simplify our formulas, we also combine C/m and ρ_0 with ε (the new ε has, again, dimensions of meter^{-1}). We thus get

$$\begin{aligned} \mathbf{f}_o &= -\varepsilon\mu \frac{\mathbf{r}'_o|\mathbf{r}'_o|}{4ar^2} e^{-\eta r} \\ &\simeq -\frac{\varepsilon\mu}{a}\left(z - (2+\eta a)\beta(1+z^2) + \cdots\right) e^{-\eta a} \end{aligned} \tag{8.17}$$

when expanded in β.

As a result we get

$$\mathcal{Q}(z) \simeq -2\varepsilon a\, \mathfrak{i} \left(1 - (2+\eta a)\frac{\beta}{z} - (1+\eta a)\beta z + \cdots\right) e^{-\eta a} \tag{8.18}$$

from which we obtain

$$a' \simeq -4\varepsilon a^2 e^{-\eta a} + \cdots \tag{8.19a}$$

$$\frac{\beta'}{\beta} \simeq -2\varepsilon a(1+\eta a)e^{-\eta a} + \cdots \tag{8.19b}$$

$$\psi' = 0 \tag{8.19c}$$

$$s'_p = 0 \tag{8.19d}$$

Since the value of ηa is typically quite large, the satellite would normally reach a near zero value of β before appreciably decreasing its altitude. But this time, even a will slowly decrease towards zero (eventually crashing into the primary).

The orbit gets distorted according to

$$\mathbf{r}_o = \mathfrak{k}\left(\frac{a(z+\beta)^2}{z(1+\beta^2)} - \frac{\varepsilon\beta a^2}{4}\left(2 + 3a\eta - 4z^2\right) e^{-\eta a}\right) \tag{8.20}$$

8.4 RADIATION PRESSURE

When the primary itself is a source of this radiation, the first term of (3.60) only modifies the value of μ in (1.27) to $\tilde{\mu} \equiv \mu - C\kappa v^2/m$ (we will assume that $\tilde{\mu}$ remains positive, which is true for particles of macroscopic size - microscopic particles are simply swept out of the solar system).

The next two terms then define the actual perturbation, called

8.4.1 Poynting-Robertson force

namely

$$\varepsilon\mathbf{f}_o = -\frac{C\kappa v}{2m}\sqrt{\frac{\tilde{\mu}}{a}}\frac{\mathbf{r}'_o}{r^3} - \frac{C\kappa v}{2m}\sqrt{\frac{\tilde{\mu}}{a}}\frac{\mathbf{r}_o\cdot\mathbf{r}'_o}{r^2}\frac{\mathbf{r}_o}{r^3} \equiv -\frac{\varepsilon}{2}\sqrt{\frac{\tilde{\mu}}{a}}\left(\frac{\mathbf{r}'_o}{r^3} + \frac{\mathbf{r}_o\cdot\mathbf{r}'_o}{r^2}\frac{\mathbf{r}_o}{r^3}\right) \tag{8.21}$$

where ε has dimensions of meter^2 sec^{-1}.

Using the first line of (5.7), this leads to

$$\mathcal{Q}(z) = \frac{-2\mathrm{i}\varepsilon(1+\beta^2)(1-\beta^2-\frac{2\beta}{z}-2\beta z)}{\sqrt{a\tilde{\mu}}(1+\frac{\beta}{z})^2(1+\beta z)^3} \tag{8.22}$$

based on (8.3b) and (8.3c).

In the manner of (5.9), (5.12) and (5.15) we get easily (the contour integrals can be evaluated by adding the corresponding function residues at $z=0$ and $z=-\beta$)

$$\frac{a'}{a} = -\frac{4\varepsilon}{\sqrt{a\tilde{\mu}}}\frac{(1+\beta^2)(1+8\beta^2+\beta^4)}{(1-\beta^2)^3} \tag{8.23a}$$

$$\frac{\beta'}{\beta} = -\frac{5\varepsilon}{\sqrt{a\tilde{\mu}}}\frac{(1+\beta^2)^2}{(1-\beta^2)^2} \tag{8.23b}$$

$$\psi' = 0 \tag{8.23c}$$

$$s_{\mathrm{p}}' = 0 \tag{8.23d}$$

and

$$\mathbf{r}_{\mathrm{o}} = \mathfrak{k}\left(\frac{a(z+\beta)^2}{z(1+\beta^2)} + \mathrm{i}\frac{3\varepsilon}{2}\sqrt{\frac{a}{\tilde{\mu}}}\beta z^2 + \mathrm{i}\frac{\varepsilon}{2}\sqrt{\frac{a}{\tilde{\mu}}}\beta^2(2z^{-1}+3z-z^3)+\cdots\right) \tag{8.24}$$

with the ε-proportional part expanded in β (the complete expression is rather lengthy — similarly to the relativistic case).

It is obvious that, to a good approximation, both a and β decrease, exponentially and at (relatively) the same rate, down to zero. This appears rather paradoxical, but one must remember that the main radiation effect (which is clearly directed outward) has been incorporated through the reduction of the μ value.

8.4.2 Distant-source radiation

We now assume that the radiation is arriving from a distant, slow moving, non-primary source (such as, for example, Sun's radiation affecting the motion of Earth's artificial satellites). In this case we get

$$\varepsilon\mathbf{f}_{\mathrm{o}} = \frac{C\kappa v^2\mathbf{R}_{\mathrm{o}}}{mR^3} \equiv \varepsilon\frac{\mathbf{R}_{\mathrm{o}}}{R^3} \tag{8.25}$$

where $\mathbf{R}$ (before being rotated to the satellite's Kepler frame) is, to a good approximation, constant; we will make it equal to $\mathrm{i}R$ by our choice of inertial coordinates. This is clearly the simplest of all perturbing forces, yet it has the most destabilizing effect on satellites' motion; in extreme cases, it can sweep a satellite out of its orbit (this happens when the relative strength of the perturbing force is bigger than about 1.3% - in that case, our technique no longer converges).

Applying the usual approach, we get

$$\mathcal{Q}(z) = \frac{2\mathrm{i}\varepsilon a^2(1+\beta z)}{\mu R^2(1+\beta^2)z}e^{-\mathrm{i}\psi}\sin\theta \tag{8.26a}$$

$$\mathcal{W}(z) = -\frac{4\mathrm{i}\varepsilon a^2\cos\theta}{\mu R^2}\left(1+\frac{\beta}{1+\beta^2}(z+\frac{1}{z})\right) \tag{8.26b}$$

leading to

$$a' = 0 \tag{8.27a}$$

$$\beta' = -\frac{3\varepsilon a^2(1+\beta^2)}{2\mu R^2}\cos\psi\sin\theta \tag{8.27b}$$

$$\phi' = \frac{6\varepsilon a^2\beta}{\mu R^2(1-\beta^2)}\sin\psi\cot\theta \tag{8.27c}$$

$$\theta' = \frac{6\varepsilon a^2\beta}{\mu R^2(1-\beta^2)}\cos\psi\cos\theta \tag{8.27d}$$

$$\psi' = \frac{3\varepsilon a^2(1-\beta^2)}{2\mu R^2\beta}\sin\psi\sin\theta - \frac{6\beta\varepsilon a^2}{\mu R^2(1-\beta^2)}\sin\psi\cos\theta\cot\theta \tag{8.27e}$$

$$s_{\mathrm{p}}' = \frac{3\varepsilon a^2(1+\beta^4)}{4\mu R^2\beta(1+\beta^2)}\sin\psi\sin\theta \tag{8.27f}$$

By numerically solving these equations, one discovers that most solutions behave as follows:

1. ψ appears to nearly flip-flop, periodically, between the value of 0 (or 2π) and π.
2. This drives the corresponding periodic oscillations of β and θ (the peaks of β coincide with $\theta \simeq 0$ or π - the orbit becoming nearly perpendicular to perturbing force). The maxima of β are often (depending on initial conditions) quite large, sometimes close to 1; the minima are always close to zero.
3. ϕ circulates by sudden systematic increases (or decreases - again, depending on initial conditions) of size π; these occur whenever ψ is leaving (but not entering) the π-value state.

For orbit's distortion, we get

$$\begin{aligned}\mathbf{r}_{\mathrm{o}} = \mathfrak{k}\Bigg(&\frac{a(z+\beta)^2}{z(1+\beta^2)} + \frac{3\varepsilon \mathfrak{i} a^2(z^2+z\beta+z^{-1}\beta^3+z^{-2}\beta^4)}{4\mu R^2(1+\beta^2)^2}e^{\mathfrak{i}\psi}\sin\theta \\ &- \frac{3\varepsilon a^2\beta(z^2\beta+z\beta^2+z^{-1}+z^{-2}\beta)}{4\mu R^2(1+\beta^2)^2}e^{-\mathfrak{i}\psi}\sin\theta\Bigg) \\ &+ \frac{\mathfrak{i}\varepsilon a^2(2-2\beta^2+2\beta^4-z^2\beta^2-z^{-2}\beta^2)}{2\mu R^2(1+\beta^2)^2}\cos\theta\end{aligned} \tag{8.28}$$

8.5 YARKOVSKY EFFECT

Based on (3.63), the Kepler-frame perturbing force is

$$\varepsilon\mathbf{f}_{\mathrm{o}} = \varepsilon e^{-\delta\mathbf{w}_{\mathrm{o}}/2} \circ \frac{\mathbf{r}_{\mathrm{o}}}{r^3} \circ e^{\delta\mathbf{w}_{\mathrm{o}}/2} \tag{8.29}$$

where

$$\varepsilon \equiv \frac{\eta C\kappa v^2}{m} \tag{8.30}$$

has dimensions of meter^3 sec^{-2}. When $\mathbf{w}$ is perpendicular to the orbital plane (implying that $\mathbf{w}_{\mathrm{o}} = \mathfrak{i}$), we get

$$\varepsilon\mathbf{f}_{\mathrm{o}} = \varepsilon\frac{\mathbf{r}_{\mathrm{o}}}{r^3} \circ e^{\mathfrak{i}\delta} = \frac{\varepsilon}{a^2}\mathfrak{k}\frac{(1+\beta^2)^2 z}{(1+\frac{\beta}{z})(1+\beta z)^3}e^{\mathfrak{i}\delta} \tag{8.31}$$

This implies

$$\mathcal{Q}(z) = \frac{2\varepsilon}{\mu}\mathfrak{k}\frac{(1+\beta^2)}{(1+\frac{\beta}{z})(1+\beta z)^2}e^{\mathfrak{i}\delta} \tag{8.32}$$

leading to

$$\frac{a'}{a} = \frac{4\varepsilon}{\mu}\frac{(1+\beta^2)^2}{(1-\beta^2)^2}\sin\delta \tag{8.33a}$$

$$\frac{\beta'}{\beta} = \frac{\varepsilon}{\mu}\frac{(1+\beta^2)^2}{1-\beta^2}\sin\delta \tag{8.33b}$$

$$\psi' = 0 \tag{8.33c}$$

$$s_{\mathrm{p}}' = \frac{\varepsilon}{2\mu}\cos\delta \tag{8.33d}$$

where the last expression represents the contribution of the $\mathbf{r}/r^3$-proportional part of the the perturbing force (previously, we dealt with such a term by correspondingly modifying μ; this is an alternate way of dealing with it).

We can see that, for positive δ, this results in a gradual increase of both a and β.

The orbit's distortion is given by

$$\mathbf{r}_{\mathrm{o}} = \mathfrak{k}\left(\frac{a(z+\beta)^2}{z(1+\beta^2)} - \mathfrak{i}\frac{\varepsilon\beta a\sin\delta}{2\mu}z^2 + \cdots\right) \tag{8.34}$$

8.6 TIDAL FORCES

We have seen in section 3.3.1 that the tidal force has the form of

$$\mathbf{f}_{\mathrm{o}} = -\varepsilon\frac{\mathbf{r}_{\mathrm{o}}(1-\delta\mathfrak{i})}{r^8} \tag{8.35}$$

where

$$\varepsilon = \frac{9}{2}mR_{\ominus}^5 \tag{8.36}$$

This translates to

$$\mathcal{Q}(z) = -\frac{2\varepsilon}{\mu a^5}\frac{(1+\beta^2)^6}{\left(1+\frac{\beta}{z}\right)^6(1+\beta z)^7}(1-\delta\mathfrak{i}) \tag{8.37}$$

which implies

$$\frac{a'}{a} = \frac{4\varepsilon\delta}{a^5\mu}\frac{(1+\beta^2)^6(1+36\beta^2+225\beta^4+400\beta^6+225\beta^8+36\beta^{10}+\beta^{12})}{(1-\beta^2)^{12}} \tag{8.38a}$$

$$\frac{\beta'}{\beta} = \frac{11\varepsilon\delta}{a^5\mu}\frac{(1+\beta^2)^7(1+10\beta^2+20\beta^4+10\beta^6+\beta^8)}{(1-\beta^2)^{11}} \tag{8.38b}$$

$$\psi' = \frac{5\varepsilon}{a^5\mu}\frac{(1+\beta^2)^6(1+10\beta^2+20\beta^4+10\beta^6+\beta^8)}{(1-\beta^2)^{10}} \tag{8.38c}$$

$$s_{\mathrm{p}}' = \frac{\varepsilon}{a^5\mu}\frac{(1+\beta^2)^5(2+17\beta^2+32\beta^4+17\beta^6+2\beta^8)}{(1-\beta^2)^9} \tag{8.38d}$$

and

$$\mathbf{r}_{\circ} = \mathfrak{k}\left(\frac{a(z+\beta)^2}{z(1+\beta^2)} - \frac{\varepsilon\beta z^2}{2a^4\mu}(5 + \mathrm{i}\delta) + \cdots\right) \tag{8.39}$$

For positive δ (such as in Earth's tidal force affecting Moon's motion), this results in a steady (but very slow) increase in a and β values.

References

[1] Arfken G, *Mathematical Methods for Physicists,* Academic Press, Orlando (1985)

[2] Beekman G: "I. O. Yarkovsky and the discovery of 'his' effect." *Journal for the History of Astronomy* **37**(2006) 71–86

[3] Berger A, Loutre M F, Laskar J: "Stability of the Astronomical Frequencies Over the Earth's History for Paleoclimate Studies" *Science* **255** (1992) 560–565

[4] Boccaletti D and Pucacco G, *Theory of Orbits, Volume 1: Integrable Systems and Non-perturbative Methods,* Springer-Verlag, Berlin (2001)

[5] Boccaletti D and Pucacco G, *Theory of Orbits, Volume 2: Perturbative and Geometrical Methods,* Springer-Verlag, Berlin (1998)

[6] Brouwer D: "Solution of the Problem of Artificial Satellite Theory without Drag" *Astronomical Journal* **64** (1959) 378–397

[7] Brown E W: "On a New Family of Periodic Orbits in the Problem of Three Bodies" *Monthly Notices of the Royal Astronomical Society* **71** (1911) 438–454

[8] Brumberg V A: "A numerical development of a generalized planetary theory" *Soviet Astronomy* **11** (1967) 156–165

[9] Brumberg V A, *Analytical Techniques of Celestial Mechanics,* Spreinger-Verlag, Berlin (1995)

[10] Coffey S L, Deprit A, and Miller B R: "The critical inclination in artificial-satellite theory" *Celestial Mechanics* **39** (1986) 365–406

[11] Cook A, *The Motion of the Moon,* Adam Hilger, Bristol and Philadelphia (1988)

[12] Danby J M A, *Fundamentals of Celestial Mechanics,* Willmann-Bell, Richmond (1992)

[13] De Moraes R V "Combined solar radiation pressure and drag effects on the orbit of artificial satellites" *Celestial mechanics* **25** (1981) 281–292

[14] Dermott S F and Murray C D: "The dynamics of tadpole and horshoe orbits 1; Theory" *Icarus* **48** (1981) 1–11

[15] Deprit A and Rom A: "The main problem of artificial satellite theory for small and moderate eccentricities" *Celestial Mechanics* **2** (1970) 166–206

[16] Doran C and Lasenby A: *Geometric Algebra for Physicists,* Cambridge University Press, Cambridge (2003)

[17] Ferraz-Mello S: "Dynamics of the asteroidal 2/1 resonance" *The Astronomical Journal* **108** (1994) 2330–2337

[18] Green R M, *Spherical Astronomy,* Cambridge University Press, Cambridge (1988)

Jan Vrbik

[19] Hamilton W R: "Memorandum respecting a new System of Roots of Unity" *Philosophical Magazine* **12** (1856) 446

[20] Hestenes D and Lounesto P: "Geometry of spinor regularization" *Celestial Mechanics* **30** (1983) 171–179

[21] Hestenes D, *New Foundations for Classical Mechanics,* Kluwer Academic Publishes, Dordrecht (1999)

[22] Kozai Y: "The Motion of a Close Earth Satellite" *Astronomical Journal* **64** (1959) 367–377

[23] Kustaanheimo P: "Spinor regularization of the Kepler motion" *Ann. Univers. Turkuensis, Ser. A* **73** (1964) 3–7

[24] Kustaanheimo P and Stiefel E L: "Perturbation theory of Kepler motion based on spinor regularization" *J. Reine Angew. Math.* **218** (1965) 204–219

[25] Milani A, Nobili A M and Knezevic Z: "Stable chaos in the asteroid belt" *Icarus* **125** (1997) 13-31

[26] Milankovitch M, *Théorie Mathématique des Phénomènes Thermiques Produits par la Radiation Solaire,* Gauthier-Villars, Paris (1920)

[27] Montenbruck O and Pfleger T, *Astronomy on the Personal Computer,* Springer-Verlag, Berlin (1994)

[28] Morbidelli Alessandro, *Modern Celestial Mechanics: Aspects of Solar System Dynamics,* Taylor & Francis, London (2002)

[29] Murray C D and Dermott S F, *Solar System Dynamics,* Cambridge University Press, Cambridge (1999)

[30] Roy A E, *Orbital Motion,* Adam Hilger, Bristol (1991)

[31] Spirig F, and Waldvogel J, *Lectures on Celestial Mechanics,* Springer-Verlag, Berlin (1985)

[32] Stiefel E L and Scheifele G, *Linear and Regular Celestial Mechanics,* Springer-Verlag, Berlin (1971)

[33] Szebehely V G, *Adventures in Celestial Mechanics: A First Course in the Theory of Orbits,* University of Texas Press, Austin (1989)

[34] Taff L G, *Celestial Mechanics: A Computational Guide for the Practitioner,* John Wiley & Sons, New York (1985)

[35] Varadi F, De la Barre, Kaula W M, and Ghil M: "Singularly weighted symplectic forms and applications to asteroid motion" *Celestial Mechanics and Dynamical Astronomy,* **62** (1995) 23-41

[36] Vivarelli M D: "The KS-transformation in hypercomplex form" *Celestial Mechanics* **29** (1983) 45–50

[37] Vivarelli M D: "The KS-transformation in hypercomplex form and the quantization of the negative-energy orbit manifold of the Kepler problem" *Celestial Mechanics* **36** (1985) 349–364

[38] Vivarelli M D:. "The Kepler problem - a unifying view" *Celestial Mechanics and Dynamical Astronomy* **60** (1994) 291–305

[39] Vivarelli M D: "The K-S transformation revisited" *Meccanica* **29** (1994) 15–26

[40] Vrbik J: "Celestial mechanics via quaternions" *Canadian Journal of Physics* **72** (1994) 141–146

[41] Vrbik J: "Two-body perturbed problem revisited" *Canadian Journal of Physics* **73** (1995) 193–198

[42] Vrbik J: "Perturbed Kepler problem in quaternionic form" *Journal of Physics* A **28** (1995) 6245–6252

[43] Vrbik J: "Resonance formation of Kirkwood gaps and asteroid clusters" *Journal of Physics* A **29** (1996) 3311–16

[44] Vrbik J: "Oblateness perturbations to fourth order" *Mothly Noticies of the Royal Astronomical Society* **291** (1997) 65–70

[45] Vrbik J: "Novel analysis of tadpole and horseshoe orbits" *Celestial Mechanics and Dynamical Astronomy* **69** (1998) 283–291

[46] Vrbik J: "Iterative solution to perturbed Kepler problem via Kustaanheimo-Stiefel equation" *Celestial Mechanics and Dynamical Astronomy* **71** (1999) 273–287

[47] Vrbik J: "Simple simulation of solar system" *Astrophysics and Space Science* **266** (1999) 557–567

[48] Vrbik J: "Perturbative solution of an asteroid's motion in resonance with Jupiter" *Monthly Notices of the Royal Astronomical Society* **316** (2000) 459–463

[49] Vrbik J: "Quaternionic processor" *Celestial Mechanics and Dynamical Astronomy* **80** (2001) 111–118

[50] Vrbik J: "Erratum: Oblateness perturbations to forth order" *Monthly Notices of the Royal Astronomical Society* **331** (2002) 1072

[51] Vrbik J: "A novel solution to Kepler's problem" *European Journal of Physics* **24** (2003) 575–583

[52] Vrbik J: "Zonal-harmonics perturbations" *Celestial Mechanics & Dynamical Astronomy* **91** (2005) 217–237

[53] Vrbik J: "Solving Lunar problem via perturbed K-S equation" *New Astronomy* **11** (March 2006) 366–373

[54] Vrbik J: "Kepler problem with time-dependent and resonant perturbations" *Journal of Mathematical Physics* **48**, 052701 (2007) 1–13

[55] Vrbik J: "Second Erratum: Oblateness perturbations to the fourth order" *Monthly Notices of the Royal Astronomical Society* **399** (2009) 1088

[56] Waldvogel J: "Quaternions and the perturbed Kepler problem" *Celestial Mechanics and Dynamical Astronomy* **95** (2006) 201–212

[57] Wisdom J: "A Perturbative Treatment of Motion near the 3/1 Commensurability" *Icarus* **63** (1985) 272–289

[58] Wnuk E.: "Tesseral harmonic perturbations for high order and degree harmonics" *Celestial Mechanics* **41** (1988) 179–191

Index

anomaly
  eccentric, 9
  mean, 9
  true, 10
apocenter, 8
attitude, 8
averaging principle, 33
axi-symmetric, 37

Cassini division, 96
center of mass, 35
chaotic behaviour, 97
complex part, 22
conjugate, 2
critical inclination, 59
cycle, *see* Lunar cycle, *see* Milankovitch cycle

drag, 42, 107

eccentricity, 8
  modfied, 8
ecliptic, 47
effect
  Poynting-Robertson, 43
  Yarkovsky, 44, 109
equation
  annual, 75
  autonomous, 18
  de-coupling, 20
  Einstein-Infeld-Hoffmann, 44
Euler angles, 3
evection, 75

force
  commensurable, 45
  perturbing, 34
  Poynting-Robertson, 107
  tidal, 41, 110
function
  analytic, 25
  hypergeometric, 48

gauge, 6
gravitational constant, 34

harmonics
  sectorial, 64
  tesseral, 57, 64
  zonal, 37, 57, 62
Hecuba gap, 90

invariant, 3

Kepler
  energy, 7
  equation, 9
  frame, 3, 18
  laws, 9
  shear, 43, 104
Kirkwood gaps, 90

Lagrange points, 81
latitude, 11
Legendre polynomials, 37
linearization, 51
longitude, 10
Lunar cycle
  nodal, 71
  perigee, 72

magnitude, 2
Milankovitch cycle
  eccentricity, 53
  obliquity, 54
  precession, 53
modified time, 5

nodal direction, 4
nutation, 73

oblateness, 38, 57
obliquity, 40
orbital elements, 16
  osculating, 33
orbits
  captured, 84
  horseshoe, 83
  hourglass, 84
  passing, 84
  tadpole, 83

primary, 34
problem
  boundary-value, 13
  initial-value, 11
  lunar, 67
  main, 57
  perturbed Kepler, 4, 14

Quaternion
  algebra, 1

Jan Vrbik

multiplication, 1

radiation pressure, 43, 107
relativistic corrections, 44, 103
resonance, 79
2/1, 90
3/1, 100
3/2, 99
5/2, 100
7/3, 101
variable, 90
retrograde, 54
rotations, 2

satellite, 34
scalar, 1
semimajor axis, 8
solar system, 45
solution
analytic, 9
iterative, 19
trial, 15
stability analysis, 82

torque, 39
Trojan asteroids, 79

variation, 75
vector, 1

www.ingramcontent.com/pod-product-compliance
Lightning Source LLC
Chambersburg PA
CBHW041718210326
41598CB00007B/699
9781608053445